To Infinity and Beyond

Eli Maor

To Infinity and Beyond

A Cultural History of the Infinite

With 162 Illustrations and 6 Color Plates

Birkhäuser
Boston · Basel · Stuttgart

Eli Maor
Associate Professor
Department of Mathematical Sciences
Oakland University
Rochester, Michigan 48063, U.S.A.

On the front jacket: Fig. 13.4. M. C. Escher: Sphere Spirals (1958). © M. C. Escher Heirs c/o Cordon Art-Baarn-Holland

Library of Congress Cataloging in Publication Data
Maor, Eli.
  To infinity and beyond.
  Bibliography; p.
  Includes index.
  1. Infinite. I. Title.
QA9.M316   1986   511.3   86–18791
ISBN 3–7643–3325–1

CIP-Kurztitelaufnahme der Deutschen Bibliothek
Maor, Eli:
To infinity and beyond: a cultural history of
the infinite / Eli Maor. — Boston; Basel;
Stuttgart: Birkhäuser, 1986.
ISBN 0–8176–3325–1 (Boston)
ISBN 3–7643–3325–1 (Basel)

ISBN 0–8176–3325–1
     3–7643–3325–1

Manufactured in the United States of America.

In memory of my teacher, Professor Franz Ollendorff
(born in Germany, 1900; died in Israel, 1981)

Preface

The infinite! No other question has ever moved so profoundly the spirit of man; no other idea has so fruitfully stimulated his intellect; yet no other concept stands in greater need of clarification than that of the infinite . . .

— David Hilbert (1862–1943)

There is a story attributed to David Hilbert, the preeminent mathematician whose quotation appears above. A man walked into a hotel late one night and asked for a room. "Sorry, we don't have any more vacancies," replied the owner, "but let's see, perhaps I can find you a room after all." Leaving his desk, the owner reluctantly awakened his guests and asked them to change their rooms: the occupant of room #1 would move to room #2, the occupant of room #2 would move to room #3, and so on until each occupant had moved one room over. To the utter astonishment of our latecomer, room #1 suddenly became vacated, and he happily moved in and settled down for the night. But a numbing thought kept him from sleep: How could it be that by merely moving the occupants from one room to another, the first room had become vacated? (Remember, *all* of the rooms were occupied when he arrived.) And then the answer dawned on our visitor: The hotel must be Hilbert's Hotel, the one hotel in town known to have an infinite number of rooms! By shifting each occupant one room over, room #1 became vacated:

This famous anecdote tells, in a way, the entire story of infinity. It is a story of intriguing paradoxes and seemingly impossible situations which have puzzled mankind for more than two millennia.

Infinity is a fathomless gulf, into which all things vanish.
□ Marcus Aurelius (121–180), Roman Emperor and philosopher

What is that thing which does not give itself, and which if it were to give itself would not exist? It is the infinite!
□ Leonardo da Vinci (1452–1519), Italian artist and engineer

The roots of these paradoxes lie in mathematics, and it is this discipline which has offered the most fruitful path towards their eventual resolution. The clarification and demystification of the infinite was fully accomplished only in our own century, and even this feat cannot be regarded as the final word. Like every science, mathematics has a refreshing air of incompleteness about it; no sooner has one mystery been solved, than a new one is already being introduced. The goal of achieving an absolute and final understanding of science is an elusive one. But it is this very elusiveness that makes the study of any scientific discipline so stimulating, and mathematics is no exception.

Many thinkers have dealt with the infinite. The philosophers of ancient Greece argued endlessly about whether a line segment—or any quantity, for that matter—is infinitely divisible, or whether an indivisible point, an "atom," would ultimately be reached. Their modern followers, the physicists, are still struggling with the same question today, using huge particle accelerators to search for the "elementary particles," those ultimate building blocks from which the entire universe is made. Astronomers have been pondering about infinity on the other extreme of the scale—the infinitely large. Is our universe infinite, as it seems to anyone watching the sky on a clear, dark night, or does it have a boundary beyond which nothing exists? The possibility of a finite universe seems to defy our very common sense, for isn't it clear that we can go forever in any direction without ever reaching the "edge"? But as we shall see, "common sense" is a very poor guide when dealing with infinity!

Artists, too, have dealt with the infinite, depicting it on canvas and in lines that became literary treasures. "I am painting the infinite," exclaimed van Gogh as he gazed at the plains of France stretching before him as far as his eyes could see. "The eternal silence of these infinite spaces terrifies me" lamented Blaise Pascal in his characteristically gloomy outlook of the world, while another man of letters, Giordano Bruno, rejoiced in the thought of an infinite universe: "Open the door through which we may look out into the limitless firmament" was his motto, for which he was arrested by the Inquisition and sentenced to die.

But however we look at the infinite, we are ultimately led back to mathematics, for it is here that the concept of infinity has its deepest roots. According to one view, mathematics *is* the science of infinity. In the *Encyclopedic Dictionary of Mathematics,* a compendium recently compiled by the Mathematical Society of Japan,[1]

---

[1] English translation published by The MIT Press, 1980.

Figure 1. *Courtesy of the Mathematical Association of America.*

THIS ISSUE : Self-reference, TNT, Shunting, Truth Machines, and Gödel's Incompleteness Theorem □ Tarr and Fether and Asylums □

the words "infinity," "infinite," and "infinitesimal" appear no fewer than 50 times in the index. Indeed, it is hard to see how mathematics could exist without the notion of infinity, for the very first thing a child learns about mathematics—how to count— is based on the tacit assumption that every integer has a successor. The notion of a straight line, so fundamental in geometry, is based on a similar assumption—that we can, at least in principle, extend a line indefinitely in both directions. Even in such seemingly "fin- ite" branches of mathematics as probability, the notion of infinity plays a subtle role: when we toss a coin ten times, we may get five "heads" and five "tails," or we may get six "heads" and four "tails," or in fact any other outcome; but when we say that the *probability* of getting "heads" or "tails" is even, we tacitly assume that an infinite number of tosses would produce an equal outcome.

My first encounter with infinity was as a young boy. I was given a book—it was the Haggadah, the story of the Exodus from Egypt—on whose cover was a picture of a young boy holding the very same book in his hand. When I looked carefully I could see the same picture on the cover of the small Haggadah the

Infinity is only a figure of speech, meaning a limit to which certain ratios may approach as closely as desired, when others are permitted to increase indefinitely.
□ Carl Friedrich Gauss (1777–1855), German mathematician

ix

boy was holding. It may even be that the picture showed up again in the picture's picture—I can't remember for sure. But I do remember that my mind was boggled by the thought that if it were possible to continue this process, it would go on forever! An intriguing possibility it was; little did I know that a relatively unknown Dutch artist, Maurits C. Escher, had been fascinated with the same idea and conveyed it in his graphic work, carrying the process to the very limits attainable with his drawing tools.

I had another encounter with infinity much later in life, an encounter of an entirely different sort. Strolling one evening along Connecticut Avenue in Washington, D.C., I found myself standing in front of a large abstract sculpture, erected right on the sidewalk. A plaque identified it as *Limits of Infinity III.* It consisted of a large elliptical bronze ring, from whose extreme points a propeller-shaped object was hinged. It looked as though the elongated object was meant to turn freely on its hinges, so I gently touched it, anticipating it to commence its motion. Instead, a hidden alarm went off, and with such a shrill sound that I was quite scared. After my initial shock was gone, I could hear an inner voice in me saying: "Thou shalt not touch infinity!"[2]

In the following chapters I have tried to share with the reader the excitement and awe that the infinite has inspired in men of all times. I took the title, *To Infinity and Beyond,* from a telescope manual that listed among the many virtues of the instrument the following: "The range of focus of your telescope is from fifteen feet to infinity and beyond." As the subtitle "A Cultural History of the Infinite" indicates, my aim is to unfold the story of infinity throughout the ages, without necessarily following a strict chronological order. My story is, to some extent, a subjective one—it is told from the point of view of a mathematician. This meant that I had to confront the same dilemma that every scientist faces when writing a book for the educated layman: How to express the author's ideas in a language understandable to the non-expert, without at the same time compromising the standards of rigor his professional peers expect of him. This dilemma is all the more

---

[2] The artist, John Safer, was kind enough to send me a most beautiful book describing his work. Referring to *Limits of Infinity III,* he says: "That turned form of bronze in the center of the piece hangs within its bronze enclosure as if floating in space. It is, the shape reminds us, the symbol of infinity." Referring to the large base that supports the work, he says: "The granite block, which is not merely a suitable resting place but a vital part of the sculpture, brings the solid and finite earth into the equation. That block of stone is *our* base from which to contemplate infinity."

Figure 2. Limits of Infinity III, *by John Safer* (*Washington, D.C.*). *Courtesy of John Safer.*

Figure 3. Infinity, *by José de Rivera (Washington, D.C.). Courtesy of the estate of José de Rivera, Grace Borgenicht Gallery, New York, and the National Museum of American History, Washington, D.C.*

acute in mathematics, which relies almost entirely on a non-verbal language of symbols and equations. I hope I have succeeded in properly addressing this problem.

As this book is intended for the general reader, I have refrained as much as possible from using "higher" mathematics in the text itself. (Of course, some familiarity with elementary algebra never does any harm.) Some specific mathematical topics have been delegated to the Appendix, thereby maintaining the continuity of the general discussion. The various chapters are, mostly, only loosely connected, so that skipping over a few of them would not impede the reading. Finally, the reader who prefers to browse through the book only casually will still enjoy the many illustrations and photographs, as well as the numerous quotations, poems, and literary lines on infinity.

Many friends helped me with this work, and I owe them many thanks. Particularly, I am indebted to my colleagues Jerrold Grossman, Wilbur Hoppe and Robert Langer, who have read large parts of the manuscript and came up with numerous suggestions; to Blagoy Trenev, whom I incessantly bothered with questions of language and style; to Hilda Bacharach and Raffaella Borasi, for bringing to my attention two beautiful poems describing the infinite; to Ruth Ollendorff, who made available to me many unpublished writings of her late husband, Professor Franz Ollendorff, to whom this book is dedicated; to Mary Besser, who edited most of the manuscript and greatly helped in its final draft; to Susan Johnson, who typed large parts of the work on the word processor; to Lynn Metzker, who prepared most of the line drawings; to the University of Wisconsin—Eau Claire and to Oakland University in Rochester, Michigan, for two generous grants that greatly helped me in my work; and finally to the editorial and production staff of Birkhäuser Boston for their special efforts to make this work a reality. But above all, I am indebted to my mother, Luise Metzger, for the intellectual enrichment she has given me over the years, and to my wife Dalia for her encouragement and patience during the many nights when I left her alone while working on the book in my office. If it were not for their support, this work would have never been completed.

Infinity is that dimension without end which the human mind cannot grasp.
□ Anonymous

One final note. At the outset of every discussion, a mathematician must define his symbols and notation. Let it therefore be known that by "he" I mean "he or she," by "him," "him or her," etc. If I am using the more traditional language in this book, it is solely for the sake of brevity.

Infinity is a floorless room without walls or ceiling.
□ Anonymous

Rochester, Michigan,
June 13, 1986

Contents

To Infinity and Beyond

Mathematical Infinity Part 1

One, two, three—
infinity

— George Gamow
(1904–1968)

1 First Steps to Infinity

There is no smallest among the small and no largest among the large; But always something still smaller and something still larger.

— Anaxagoras (*ca.* 500–428 B.C.)

Infinity has many faces. The layman often perceives it as a kind of "number" larger than all numbers. For some primitive tribes infinity begins at three, for anything larger is "many" and therefore uncountable. The photographer's infinity begins at thirty feet from the lens of his camera, while for the astronomer—or should I say the cosmologist—the entire universe may not be large enough to encompass infinity, for it is not at present known whether our universe is "open" or "closed," bounded or unbounded. The artist has his own image of the infinite, sometimes conceiving it, as van Gogh did, as a vast, unending plane on which his imagination is given free rein, at other times as the endless repetition of a single basic motif, as in the abstract designs of the Moors. And then there is the philosopher, whose infinity is eternity, divinity, or the Almighty Himself. But above all, infinity is the mathematician's realm, for it is in mathematics that the concept has its deepest roots, where it has been shaped and reshaped innumerable times, and where it finally celebrated its greatest triumph.

Mathematical infinity begins with the Greeks. To be sure, mathematics as a science had already reached quite an advanced stage long before the Greek era, as is clear from such works as the Rhind papyrus, a collection of 84 mathematical problems written in hieratic script and dating back to 1650 B.C.[1] But the ancient

---

[1] The papyrus is named after the Scottish Egyptologist A. Henry Rhind, who purchased it in 1858. It is now in the British Museum. See *The Rhind Mathematical Papyrus* by Arnold Buffum Chace, The National Council of Teachers of Mathematics, Reston, Virginia, 1979.

mathematics of the Hindus, the Chinese, the Babylonians, and the Egyptians confined itself solely to practical problems of daily life, such as the measurement of area, volume, weight, and time. In such a system there was no place for as lofty a concept as infinity, for nothing in our daily lives has to deal directly with the infinite. Infinity had to wait until mathematics would make the transition from a strictly practical discipline to an intellectual one, where knowledge for its own sake became the main goal. This transition took place in Greece around the sixth century B.C., and it thus befell the Greeks to be the first to acknowledge the existence of infinity as a central issue in mathematics.

Acknowledge—yes, but not confront! The Greeks came very close to accepting the infinite into their mathematical system, and they just might have preceded the invention of the calculus by some two thousand years, were it not for their lack of a proper system of notation. The Greeks were masters of geometry, and virtually all of classical geometry—the one we learn in school— was formulated by them. Moreover, it was the Greeks who introduced into mathematics the high standards of rigor that have since become the trademark of the profession. They insisted that nothing should be accepted into the body of mathematical knowledge that could not be logically deduced from previously established facts. It is this insistence on *proof* that is unique to mathematics and distinguishes it from all other sciences. But while the Greeks excelled in geometry and brought it to perfection, their contribution to algebra was very meager. Algebra is essentially a language, a collection of symbols and a set of rules by which to operate with these symbols (just as a spoken language consists of words and of rules by which to combine these words into meaningful sentences). The Greeks did not possess the algebraic language, and consequently were deprived of its main advantages—the generality it offers and its ability to express in an abstract way relations between variable quantities. It is this fact, more than anything else, which brought about their *horror infiniti,* their deeply rooted suspicion of the infinite. "The infinite was taboo," said Tobias Dantzig in his classic work, *Number—the Language of Science,* "it had to be kept out, at any cost; or, failing this, camouflaged by arguments *ad absurdum* and the like."

Nowhere was this fear of the infinite better manifested than in the famous paradoxes of Zeno, a philosopher who lived in Elea in the fourth century B.C. His paradoxes, or "arguments," as they were called, deal with motion and continuity, and in one of them he proposed to show that motion is impossible. His argument seems quite convincing: in order for a runner to move from

3

Figure 1.1. *The runner's paradox.*

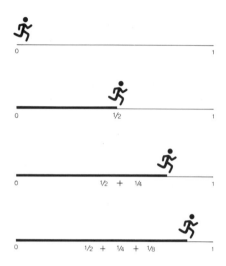

one point to another, he must first cover half the distance between the two points, then half of the remaining distance, then half of what remains next, and so on *ad infinitum* (Fig. 1.1) Since this requires an infinite number of steps, Zeno argued, the runner would never reach his destination. Of course, Zeno knew full well that the runner would reach the end point after a finite lapse of time. Yet he did not resolve the paradox; rather, he left it for future generations. In this at least he was humble, admitting that the infinite was beyond his and his generation's intellectual reach. Zeno's paradoxes had to wait another twenty centuries before they would be resolved.

But while the Greeks were unable to grasp the infinite intellectually, they still made some good use of it. They were the first to devise a mathematical method to find the value of that celebrated number which we denote today by π, the ratio of the circumference of a circle to its diameter. This number has intrigued laymen and scholars alike since the dawn of recorded history. In the Rhind Papyrus (*ca.* 1650 B.C.), we find its value to be $(4/3)^4$, or very nearly 3.16049, which is within 0.6% of the exact value. It is indeed remarkable that the ancient Egyptians already possessed such a degree of accuracy. The Biblical value of π, by comparison, is *exactly 3,* as is clear from a verse in 1 Kings vii 23: "and he made a molten sea, ten cubits from brim to brim, and his height was five cubits; and a line of thirty cubits did encompass him round about." Thus the error in the Biblical value is more than 4.5% of the true value of π!

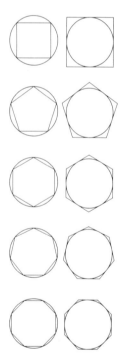

Figure 1.2. *Regular inscribed and circumscribing polygons.*

Now all the ancient estimations of π were essentially empirical—

4

they were based on an actual measurement of the circumference and diameter of a circle. The Greeks were the first to propose a method which would give the value of π to any degree of accuracy by a *mathematical* process, rather than by measurement. The inventor of this method was Archimedes of Syracuse (*ca.* 287–212 B.C.), the great scientist who achieved immortal fame as the discoverer of the laws of floating bodies and of the mechanical lever. His method was based on a simple observation: take a circle and circumscribe it by a series of regular polygons of more and more sides. (In a regular polygon, all the sides and angles are equal.) Each polygon has a perimeter slightly in excess of the circumference of the circle; but as we increase the number of sides, the corresponding polygons will encompass the circle more and more tightly (Fig. 1.2). Thus if we can find the perimeters of these polygons and divide them by the diameter of the circle, we will get a fairly close approximation to π. Archimedes followed this procedure for polygons having 6, 12, 24, 48, and 96 sides, for which he was able to calculate the perimeters by using known methods. For the 96-sided polygon he found the value 3.14271 (this is very close to the $^{22}\!/_{7}$ approximation often used in school). He then repeated the same procedure with polygons touching the circle from within—inscribed polygons—giving values that are all short of the true value. Again by using 96 sides, Archimedes found the value 3.14103, or very nearly $3^{10}\!/_{71}$. Since the actual circle is "squeezed" between the inscribed and circumscribed polygons, the true value of π must be somewhere between these values.

Let there be no misunderstanding: Archimedes' method gave an estimation of π far better than anything before him. But its real innovation was not in this improved value, but in the fact that it enabled one to approximate π *to any desired accuracy,* simply by taking polygons of more and more sides. In principle, there is no limit to the degree of accuracy this method could yield—even though for practical purposes (such as in engineering), the above values are more than adequate. In modern language we say that π is the *limit* of the values derived from these polygons as the number of sides tends to infinity. Archimedes, of course, did not mention the limit concept explicitly—to do so would have required him to use the language of algebra—but two thousand years later this concept would be the cornerstone around which the calculus would be erected.

n	inscribed n-gon	superscribed n-gon
3	2.59808	5.19615
6	3.00000	3.46410
12	3.10583	3.21539
24	3.13263	3.15966
48	3.13935	3.14609
96	3.14103	3.14271
192	3.14145	3.14187

Approximations for π using different inscribed and circumscribed polygons. n denotes the number of sides in each case.

$\pi \approx 3.1415926535$
8979323846
2643383279
5028841971
6939937510
5820974944
5923078164
0628620899
8628034825
3421170679

π approximated to one hundred decimal places.

Zero, One, Infinity

יִגְדַּל אֱלֹהִים חַי וְיִשְׁתַּבַּח   נִמְצָא וְאֵין עֵת אֶל מְצִיאוּתוֹ:

אֶחָד וְאֵין יָחִיד כְּיִחוּדוֹ   נֶעְלָם וְגַם אֵין סוֹף לְאַחְדּוּתוֹ:

The living God 0 magnify and bless,
Transcending time and here eternally.
One being, yet unique in unity,
A mystery of Oneness, measureless.

— from a prayer based on Maimonides'
Thirteen Principles of the Jewish Faith

Zero and infinity are often perceived by the layman as synonyms for "nothing" and "numerous." The notion that an endless division results in zero is quite common, and one can still find the equations

$$\frac{1}{0} = \infty \quad \text{and} \quad \frac{1}{\infty} = 0$$

in many older textbooks. However, except for one special case which we will discuss in Chapter 12, these equations are meaningless. Zero is a number, an integer like all other integers (albeit one with a unique role). Infinity, on the other hand, is a *concept;* it is not part of the real number system and cannot therefore be related to a real number in the same sense that numerical quantities can.

The word *zero* originated from the Hindu *sunya,* which means "empty" or "blank"; it was used by the Hindus as early as the ninth century A.D. to indicate an empty slot in their positional system of numeration, so that one could distinguish, say, between 12 and 102. When the Arabs imported the Hindu system to Europe, the *sunya* was translated into the Arabic *as-sifr,* which was later transliterated into the Latin *zephirum.* This in turn became the Italian *zeuero,* from which it was but one more step to our modern zero. The symbol 0 first appears in a Hindu inscription from the year 870 A.D.

Actually the symbol 0 has several meanings in mathematics. On the number line, 0 stands not only for the number 0 but also for the *position* associated with this number—the starting point from which we both count and move. In the plane, the point 0 denotes the origin of the coordinate system, the point whose coordinates are (0,0). A set without any elements—the *empty* or *null set*—is denoted by the Greek letter ϕ (phi), a symbol reminiscent of the ordinary 0. In higher algebra one deals with *matrices*—arrays of numbers arranged in rows and columns. One can add, subtract, multiply, and in a certain sense divide these matrices. Thus if **A** is the matrix $\begin{pmatrix} 2 & 5 & 8 \\ 3 & 0 & -1 \end{pmatrix}$ and **B** is the matrix $\begin{pmatrix} 7 & 1 & 3 \\ 2 & -4 & 6 \end{pmatrix}$, then **A** + **B** is the matrix $\begin{pmatrix} 9 & 6 & 11 \\ 5 & -4 & 5 \end{pmatrix}$ obtained by adding the corresponding elements of **A** and **B**. A matrix whose entries are all zeros is called the zero matrix and is denoted by **O**. It shares many of the properties of the number 0; for example **A** + **O** = **A**. Such matrices are very useful in a host of situations, among them the analysis of mechanical vibrations, electric circuits, and economic systems.

Whatever the context in which it is used, division by zero is meaningless. For, suppose we attempt to divide 5 by 0. Call the outcome x: $x = 5/0$. The number x, by definition, must fulfill the equivalent equation $0 \cdot x = 5$ (just as the equations $6/2 = 3$ and $2 \cdot 3 = 6$ are equivalent). But $0 \cdot x$ is always equal to 0, leading to the absurd result $0 = 5$. Thus no number x can satisfy the equation $x = 5/0$. But what about dividing 0 by 0? Again put $x = 0/0$, so that $0 \cdot x = 0$. Now this equation is satisfied by *every* number x, so that we don't get a definite, unique answer. (Similar complications arise when one attempts to "divide" a matrix by the zero matrix.) Either situation is unacceptable mathematically, and division by zero is therefore declared to be an invalid operation.

While an expression such as $1/0$ is meaningless, it is true that when one divides a number, say 1, by smaller and smaller divisors, the result becomes larger and larger. We say that the *limit* of $1/x$ as x tends to zero is infinite, which is just another way of saying that $1/x$ grows without bound as x tends to 0. The situation is summarized by the equation $\lim_{x \to 0} 1/x = \infty$.

While the symbol 0 has been in use for over a thousand years, that for infinity, ∞, is of a more recent vintage. It was first used in 1655 by the English mathematician John Wallis (1616–1703), who was also a classical scholar and probably took it from the Roman numeral for 100 million (Fig. 1).

The number 0 and the concept of infinity are indispensable in mathematics, but their role would not be complete without a

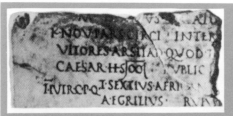

Figure 1. *The Roman numeral for 100 million consisted of the infinity symbol ∞ placed within a frame. From an inscription dating from the year 36 A.D. Reprinted from Karl Menninger,* Number Words and Number Symbols—A Cultural History of Numbers, *The MIT Press, 1977, with permission of Vandenhoeck & Ruprecht, Göttingen.*

Figure 2. *Infinity symbol on a camera lens. Courtesy of Minolta Corporation.*

Figure 3. *Trademark of Dr. Hahn, Germany. Reprinted from Yusaka Kamekura,* Trademark Designs, *Dover Publications, New York, 1980, with permission.*

third element—the number one. Mathematically, this number is the generator from which all positive integers are formed by successive addition: $2 = 1 + 1$, $3 = 2 + 1$, $4 = 3 + 1$, etc. (The negative integers can be similarly formed by successive subtraction.) Geometrically, each of the symbols 0, 1, and ∞ has a unique role on the number line: 0 represents the starting point, 1 the scale we use (i.e., the size of the unit), and ∞ the completeness of the line—the fact that it includes *all* real numbers.

The number one has often been given philosophical and even divine attributes; in most religions it symbolizes the oneness of God, and as such has often been equated with the infinite, the omnipotence of the Creator. Aristotle refused to regard one as a number, since it is the generator of all numbers but cannot itself be generated. This view was held by many thinkers well into the Middle Ages, but of course such interpretations have very little to do with mathematics.

On the other hand, in higher mathematics we often encounter the so-called "indeterminate forms," expressions such as ∞/∞. Such an expression has no preassigned value; it can only be evaluated through a limiting process. Consider, for example, the expression $(2x + 1)/(x - 1)$. As x tends to infinity, both the numerator and denominator grow without bound, yet their ratio approaches the limit 2: $\lim_{x \to \infty} (2x + 1)/(x - 1) = 2$. It would be totally incorrect, though, to write $\infty/\infty = 2$; in fact, had we considered the expression $(2x + 1)/(3x - 1)$ instead, the limit would have been $2/3$.

Another indeterminate form is $\infty - \infty$. It would be tempting to say that since any number when subtracted from itself gives 0, we should have $\infty - \infty = 0$. That this may be false can be seen from the expression $1/x^2 - (\cos x)/x^2$, where $\cos x$ is the cosine function defined in trigonometry. As $x \to 0$, each term tends to infinity, yet it can be shown with some effort that the

8

entire expression tends to the limit $1/2$: $\lim_{x \to 0} [1/x^2 - (\cos x)/x^2] = 1/2$. Loosely speaking, in each indeterminate expression there is a "struggle" between two quantities, one tending to make the expression numerically large, the other tending to make it small. The final outcome depends on the precise limiting process involved. The seven indeterminate forms encountered most frequently in mathematics are $0/0$, ∞/∞, $\infty \cdot 0$, 1^∞, 0^0, ∞^0, and $\infty - \infty$.

Figure 4. *Infinity symbol on a Swiss stamp, "Time and Eternity." With permission of the PTT, Switzerland.*

2 Towards Legitimation

It is only the affirmation of the power of the mind which knows it can conceive of the indefinite repetition of the same act, when the act is once possible.

— Henri Poincaré (1854–1912)

Like most other sciences, European mathematics came to a virtual standstill during the long, dark Middle Ages. It was not until the sixteenth century that the notion of infinity—long since forgotten as a scientific issue and having become instead the subject of theological speculations—underwent its revival. And one of the first questions to be tackled was once again that of finding an approximation to the value of π. The result was a remarkable formula, one which even today arouses our admiration for its beauty:

$$\frac{2}{\pi} = \frac{\sqrt{2}}{2} \cdot \frac{\sqrt{2+\sqrt{2}}}{2} \cdot \frac{\sqrt{2+\sqrt{2+\sqrt{2}}}}{2} \cdots$$

This *infinite product* was discovered in 1593 by the French mathematician François Viète (1540–1603); it shows that π can be calculated solely from the number 2 by a succession of additions, multiplications, divisions, and square root extractions—that is, by the elementary operations of school mathematics. But the most important feature of the formula is the three dots at the end, which tell us to go on, and on, and on, . . . , *ad infinitum*. This was the first time that an infinite process was explicitly expressed as a mathematical formula, and it heralded the beginning of a new era. No longer would the infinite be something ominous, some vague concept that should be avoided at all costs; quite to the contrary, it could now be put in writing and thus be legitimately accepted into the kingdom of mathematics.

Soon to follow were other formulas that required the infinite application of the basic arithmetic operations, formulas which were

10

no less remarkable than Viète's. One of these was discovered in 1650 by the English mathematician John Wallis (1616–1703), and once again it involved the number π:

$$\frac{\pi}{2} = \frac{2 \cdot 2 \cdot 4 \cdot 4 \cdot 6 \cdot 6 \cdots}{1 \cdot 3 \cdot 3 \cdot 5 \cdot 5 \cdot 7 \cdots}$$

(It was Wallis, incidentally, who proposed the symbol ∞ for infinity.) And in 1671, the Scotch James Gregory (1638–1675) found yet another formula involving π, this one an *infinite series:*

$$\pi/4 = 1/1 - 1/3 + 1/5 - 1/7 + - \cdots$$

(This series was discovered independently in 1674 by Gottfried Wilhelm Leibniz, co-inventor with Newton of the calculus, and is sometimes referred to as the Gregory–Leibniz series.) We will have more to say about this series and similar ones later on.

The essence of these formulas is that by calculating more and more of their terms, the value of π can be found, at least in principle, to as many decimal places as we wish. Today we know that the very nature of the number π is such that we can never find its "exact" value, because such a knowledge would require an infinite number of digits. To be sure, modern computers have calculated its value to millions of digits—a process which is totally useless from a practical point of view, but which nevertheless is of some theoretical interest, as it enables one to examine the statistical distribution of the digits in the decimal expansion and look for any possible patterns, should they exist (so far none have been found). Numbers such as π and e (the base of natural logarithms, whose approximate value is 2.71828) belong to a special class of numbers called *transcendental* by mathematicians—a word which, in spite of its mystical flavor, has a very precise meaning in mathematics.[1] The transcendence of π was not proved until 1882, and with this came to a conclusion almost four millennia of search into the nature of this unique number.

Simultaneously with these developments, the infinite also enjoyed a renaissance of a different kind. As we have seen, Archimedes' method of approximating π came very close to our modern differential and integral calculus. This, however, was only one of his contributions to mathematics. He was particularly interested

---

[1] A number is called *algebraic* if it is a solution of an algebraic equation, i.e., a polynomial equation whose coefficients are integers (whole numbers). Thus the numbers 5, $-2/3$, $\sqrt{2}$, and $2 + \sqrt{3}$ are all algebraic, because they are solutions of the equations $x - 5 = 0$, $3x + 2 = 0$, $x^2 - 2 = 0$, and $x^2 - 4x + 1 = 0$, respectively. A number is *transcendental* if it is not algebraic; that is, if it is not a solution of any algebraic equation.

11

Figure 2.1. *Area under a parabola.*

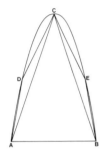

Figure 2.2. *Area under a parabola: the method of exhaustion.*

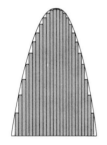

Figure 2.3. *Area under a parabola: the method of indivisibles.*

in finding the areas and volumes of various plane figures and solids, and was the first to find the area of a segment of the parabola, the curve traced by a flying projectile in the absence of air resistance (Fig. 2.1). He did this by the "method of exhaustion"—approximating the figure (or solid) by a series of small parts whose areas (or volumes) can be found, and then adding them up to obtain the desired quantity (Fig. 2.2). The idea behind the method was that by making these "elementary" parts smaller and smaller, one could make them fit the figure as closely as one pleased. Thus an infinite process was subtly involved, but Archimedes carefully avoided any direct reference to it. During the first half of the seventeenth century there was a renewed interest in this method, inspired no doubt by the new discoveries in astronomy which showed that the planets and comets move around the sun in elliptical and parabolic orbits. But the modern forefathers of the calculus had no patience for the ancient Greek standards of rigor; applications to science and engineering were the call of the day, and so these men used Archimedes' method to their advantage while avoiding its pedantic subtleties. The result was a crude contraption that had none of the elegance of the Greek methods but which somehow seemed to work and produce results—the "method of indivisibles."[2] Using this method, mathematicians like Galileo Galilei (1564–1642), Johannes Kepler (1571–1630), and Bonaventura Cavalieri (1598–1647) discovered numerous properties of the various figures and solids of common geometry, as well as a host of applications in mechanics and optics based on them. Cavalieri, in particular, was the driving force in promoting the method of indivisibles which he did in his book *Geometria indivisibilibus continuorum,* published in 1635. In using the "indivisibles" and perceiving them as the "atoms" of area or volume, these scientists came within one step of our modern integral calculus. (An example of this method is shown in Fig. 2.3)

It was with Sir Isaac Newton (1642–1727) and Gottfried Wilhelm Leibniz (1646–1716), during the second half of the seventeenth century, that the infinite celebrated its greatest triumph thus far, for it was the key element in their newly invented differen-

---

[2] The main difference between the method of exhaustion and the method of indivisibles is that the former relies only on finite quantities (albeit of diminishing size), while the latter regards a shape as being made up of infinitely many elements, each infinitely small. Although the method of exhaustion has a sounder mathematical foundation, both methods were but a disguise for using the limiting process without explicitly admitting it.

tial and integral calculus. It all revolved around the infinitely small, the *infinitesimal*, as it came to be known. Much controversy was stirred by these whimsical creatures, the infinitesimals. Were they finite quantities? If so, why not regard them as ordinary, down-to-earth numbers? Or were they vanishing quantities, actually equal to zero? Then why use them in the first place? No one has expressed these doubts more pointedly than Bishop George Berkeley (1685–1753), an English philosopher and theologian, who in 1734 published a satirical work entitled *The Analyst, Or a Discourse Addressed to an Infidel Mathematician*, [3] in which he attacked the new calculus on various grounds—including theological ones—and ridiculed its very foundations. Speaking of the infinitesimals he said:

> *They are neither finite quantities, nor quantities infinitely small, nor yet nothing. May we not call them the ghosts of departed quantities?*

Still, the fact remained that numerous problems in mathematics, physics, and astronomy could suddenly be solved with the new methods of the calculus—and solved very efficiently, for that matter. It thus happened that what had first been severely criticized by the "pure" mathematicians as an illegitimate child of incorrect reasoning was soon legalized and adopted by their "applied" colleagues, the physicists, astronomers, and engineers. The full story of this singular event, with its accompanying personal drama, is far too long to be told here in detail. (Newton and Leibniz made their discoveries independently of each other and from somewhat different points of view, precipitating a bitter priority dispute that was not without some political undertones, involving the prestige of their respective countries, Britain and Germany.) Suffice it to say that most of our present knowledge of science and technology would have been unthinkable without the calculus. And the calculus, in turn, opened the way to a vast and fruitful branch of mathematics known as *analysis*, embracing virtually everything that involves continuity and change and, therefore, the infinite.

---

[3] The infidel was the English astronomer Edmond Halley (1656–1742), who is best known for studying the comet that bears his name but who was also a mathematician of considerable talent. Halley helped Newton publish his major work, the *Principia* (1687).

Numbers Large and Small

Laymen—and children especially—are always fascinated by large numbers. To the mathematician, however, they have no particular significance; in fact, the two most important constants of mathematics, π and e, have quite "ordinary" values (about 3.14 and 2.72, respectively). It is true that the fundamental constants of physics usually have very large or very small values (for example, the speed of light is 3×10^{10} cm/sec, the mass of the electron is 9.1×10^{-28} gm, and Planck's constant is 6.6×10^{-27} erg sec). But this is so only because these constants are expressed in terms of our ordinary units of measurement, which are derived from the dimensions of our earth. Thus, one centimeter is one four-billionth of the length of the equator, one gram is the mass of one cubic centimeter of water at $0°C$, and one second is one part in 86,400 of the length of a solar day. Had we used other units—such as the known radius of the universe or the time since its creation—the fundamental constants would have had quite different numerical values.

To facilitate the writing of very large and very small numbers, scientists use the so-called "scientific notation," which employs powers of 10. Thus, 1,000,000 (one million) is written as 10^6, 1,000,000,000 (one billion) as 10^9, and so on. Very small numbers are written with negative exponents: 0.000,001 (one millionth) is 10^{-6}, 0.000,000,001 (one billionth) is 10^{-9}, etc. The number 123,000 is written as 1.23×10^5 while 0.00123 is written as 1.23×10^{-3}.

Very large numbers appear quite frequently in problems involving combinations and permutations. For example, there are $1 \times 2 \times 3 = 6$ different ways of arranging three objects in a row

14

Table 1. *Values of* n-*factorial.*

n	$n!$
0	1 (by definition)
1	1
2	2
3	6
4	24
5	120
6	720
7	5040
8	40,320
9	362,880
10	3,628,800
11	39,916,800
12	479,001,600
13	6,227,020,800
14	87,178,291,200
15	1,307,674,368,000
16	20,922,789,888,000
17	355,687,428,096,000
18	6,402,373,705,728,000
19	121,645,100,408,832,000
20	2,432,902,008,176,640,000
21	51,090,942,171,709,440,000
22	1,124,000,727,777,607,680,000
23	25,852,016,738,892,566,840,000
24	620,448,401,733,421,599,360,000
25	15,511,210,043,335,539,984,000,000
26	403,291,461,126,724,039,584,000,000
27	10,888,869,450,421,549,068,768,000,000
28	304,888,344,611,803,373,925,504,000,000
29	8,841,761,993,742,297,843,839,616,000,000
30	265,252,859,812,268,935,315,188,480,000,000
40	8.15915×10^{47} (approximately)
50	3.04141×10^{64} (approximately)
60	8.32099×10^{81} (approximately)
70	1.19786×10^{100} (approximately)
80	7.15695×10^{118} (approximately)
90	1.48572×10^{138} (approximately)
100	9.33262×10^{157} (approximately)

(ABC, BCA, CAB, ACB, BAC, and CBA). For four objects there are $1 \times 2 \times 3 \times 4 = 24$ different ways, and in general for n objects there are $1 \times 2 \times 3 \times \cdots \times n$ different ways. This number is called "n-factorial" and written $n!$. It grows very fast with increasing n, as Table 1 shows.

In the famous Rubik Cube, there are 43,252,003,274,489,856,000 different ways of arranging its six faces, or about 4×10^{19}—far more than the "three billion combinations" that some of the cube's manufacturers boast about!

The ancients seem to have had a mythical reverence for large numbers, which, to them, symbolized power and fertility. Thus God blessed Abram: "Look now toward heaven, and count the stars, if thou be able to count them . . . So shall thy seed be" (Genesis 15:5). He then went on: "Neither shall thy name any more be called Abram, but thy name shall be Abraham; for the father of a multitude of nations have I made thee" (Genesis 17:5; the "h" comes from the Hebrew word for multitude, *ha'mon*). Ironically, the stars at night are not as numerous as one might think. Under ideal conditions the naked eye can see only some 2,800 stars from any given point on the earth at any given time. The sand grains on a beach are a much more appropriate symbol of plenty. However, Archimedes, in his *The Sand-Reckoner*, demonstrated that even this huge number can be counted:

> *Many people believe, King Gelon, that the grains of sand are without number. Others think that although their number is not without limit, no number can ever be named which will be greater than the number of grains of sand. But I shall try to prove to you that among the numbers which I have named there are those which exceed the number of grains in a heap of sand the size not only of the earth, but even of the universe.*

The largest number with an independent name outside the decimal system of nomenclature is the Buddhist *asankhyeya,* equal to 10^{140}. The largest named number in the Western world is the *Googol,* or 10^{100}. The name was given by the American mathematician Edward Kasner—or rather by his nine-year-old nephew, whom Kasner had asked to invent a name for a very large number. Ten raised to the power of one Googol is a *Googolplex:* 1 followed by a Googol zeros. Such monstrous numbers are purely intellectual creations and are rarely, if ever, used anywhere. For example, the total number of atoms in the entire universe is estimated at "only" 10^{85}—far less than one Googol.

Big as these numbers are, they have nothing to do with infinity. In fact, infinity is as remote from a Googol as from 1. A variable quantity is said to approach infinity if it can become *larger than any finite number,* no matter how large. It follows that infinity is not a number at all, but a concept.

Convergence and Limit [']   3

*A quantity is the limit of another quantity, when the second can
approach the first more closely than any given quantity as small
as one can suppose.*

— Jean Le Rond D'Alembert (1717–1783)

Central to the development of the calculus were the concepts of
convergence and limit, and with these concepts at hand it became
at last possible to resolve the ancient paradoxes of infinity which
had so much intrigued Zeno. For example, the runner's paradox
is explained by the following observation: By first covering one-
half the distance between the runner's starting and end points,
then half the remaining distance, and so on, he will cover a total
distance equal to the sum:

$$1/2 + 1/4 + 1/8 + 1/16 + \cdots$$

This infinite sum, or series, has the property that no matter how
many of its terms we add up, we will never reach 1, let alone
exceed 1; and yet, we can make the sum get as close to 1 as we
please, simply by adding a sufficiently large number of terms (Fig.
3.1). We say that the series *converges* to 1, or that it has the number

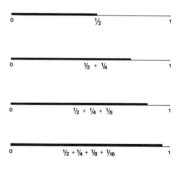

Figure 3.1. *The runner's
paradox revisited.*

17

1 as its *limit,* as the number of added terms increases to infinity. Now, assuming that the runner maintains a steady speed, the time intervals that it takes him to cover these distances will also follow the same series; therefore, he will cover the entire distance in a finite span of time—which settles the issue.[1] What the Greeks had refused to accept was the fact that an infinite sum may add up to a finite value, that is, may converge to a limit.

Before we can take a closer look at infinite series, we must explain what is meant by the limit of an infinite *sequence,* or *progression.* A sequence is simply a row of numbers, written as a_1, a_2, a_3, . . . , a_n, . . . , where there is usually (though not always) some rule that tells us how to obtain the next number in the sequence. The notation a_n stands for "the nth term," or "nth member," of our sequence, and the three dots following it remind us—lest we have forgotten!—that the sequence will never come to an end. (There would hardly be any interest in studying the limit of a finite sequence.)

Now it may happen that if we go far enough along a given sequence, its terms will approach closer and closer to a definite number, without ever reaching it. If this happens, that number is called the *limit* of our sequence, and we say that the sequence *converges* to this limit. The sequence 1, 1/2, 1/3, 1/4, . . . , 1/n, . . . , for example, has the number zero as its limit. The farther we go along this sequence, the smaller its terms become; they *approach* zero, but never actually become zero. In fact, we can make the members of our sequence get as close to zero as we please—all we have to do is go far enough. Thus, if we want our members to become smaller than, say, one-thousandth, we can easily achieve this by going at least as far as the thousandth term ($n = 1,000$); if we want them to become smaller than one-millionth, we will have to go at least until the millionth term ($n = 1,000,000$); if this is not close enough, we can make our terms approach zero within one-billionth, one-trillionth, and so on—always by going "far enough out." This is precisely the essence of the limit concept: the limit is a number that a sequence can approach as closely as we please, but never actually reach.[2]

---

[1] Notwithstanding this, there are some who, even today, reject this simple explanation and refuse to regard the matter as settled. Apparently there will always be those who stubbornly reject any argument that demands an insight beyond our immediate "common sense." To this group belong the circle-squarers, the flat-earth believers, and those who, despite all existing evidence, will take up any opportunity to denounce the theory of relativity.

[2] We exclude the trivial case where all the members of the sequence

Figure 3.2. *Convergence to a limit.*

Pictorially, this means that the members of the sequence will crowd together near their limiting value (zero in our example), as in Fig. 3.2.

Now mathematicians loathe long, verbal explanations; they are in the habit of making their statements brief and concise. Instead of saying that the sequence $a_1, a_2, a_3, \ldots, a_n, \ldots$ converges to the limit L, we write:

$$a_n \to L \text{ as } n \to \infty$$

or alternatively

$$\lim_{n \to \infty} a = L$$

However, the equality sign in the second notation needs a word of caution: all it says is that the *limit* of the sequence is L; it does not say—and it should not be inferred—that this limit is actually reached. Thus, for the sequence $1, 1/2, 1/3, \ldots, 1/n, \ldots$, the fact that the terms approach the limit zero would be written as $1/n \to 0$ *as* $n \to \infty$, or $\lim_{n \to \infty} 1/n = 0$.

The definition of limit allows for considerable freedom as to the exact manner in which the limit is approached. In the above example, the members of the sequence approach their limit (zero) from one direction, namely, from above (or from the right on the number line); that is, the members are always greater than zero. But this need not be the case. Take, for example, the same sequence but with alternating signs: $1, -1/2, 1/3, -1/4, \ldots$.[3] This sequence also has the limit zero, but this time the limit is approached alternately from above and from below (i.e., from the right and from the left on the number line) in an oscillatory manner (Fig. 3.3). Nor need the convergence be "monotonic" (i.e., such that each additional term brings us closer to the limit).

Figure 3.3. *Convergence to a limit: an oscillating sequence.*

---

are equal, or where the limiting value itself is "inserted" as an isolated member into the sequence. The requirement of getting arbitrarily close to the limit will, of course, cover these cases too.

[3] The general term of this sequence can be written as $\dfrac{(-1)^{n-1}}{n}$.

The sequence 2/1, 1/2, 3/2, 2/3, 4/3, 3/4, . . . converges to the limit 1 in "jumps," in addition to being oscillatory. The sole requirement is that the limit can be approached arbitrarily closely.

The Prime Numbers

The primes . . . . those exasperating, unruly integers that refuse to be divided evenly by any integers except themselves and 1.

— Martin Gardner

An integer such as 12 can be written as a product of smaller integers: $12 = 3 \times 4$ or $12 = 2 \times 6$. Such a number is called *composite*. But the integer 13 can only be written as a product of itself and 1: $13 = 13 \times 1$. Such a number is called *prime*. More precisely, a prime number is a positive integer greater than 1 that cannot be evenly divided by any other integer except itself and 1.

Except for 0 and 1, every natural number is either composite or prime. The first ten primes are 2, 3, 5, 7, 11, 13, 17, 19, 23, and 29. The numbers 0 and 1 are regarded as neither prime nor composite. All primes except 2 are odd numbers, since an even number is divisible by 2 and is therefore composite.

The prime numbers play a fundamental role in mathematics, particularly in higher arithmetic (the branch of mathematics known as "number theory"). The reason is that every composite number can be factored into primes in one and only one way. Thus 12 can be factored into (i.e., written as a product of) 3 and 4 or into 2 and 6, but $4 = 2 \times 2$ so that $12 = 3 \times 4 = 3 \times 2 \times 2$, or again $6 = 2 \times 3$ so that $12 = 2 \times 6 = 2 \times 2 \times 3$. Thus, except for their order, we arrive at the same prime factors. This fact, known as the *Fundamental Theorem of Arithmetic*, shows that the primes are the "building blocks" of all numbers. They play the same role in mathematics as the chemical elements in the physical world.

However, there are two facts which make this analogy incomplete. First, unlike the chemical elements, there are infinitely many primes. Secondly, there is no "periodic table" of primes, no apparent regular pattern into which they fit.

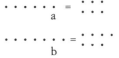

Figure 1. (*a*). *A composite number such as six can be represented by beads arranged in rows, each row having the same number of beads (6 = 2 × 3). (b). A prime number such as seven cannot be represented in this manner: a remainder of one bead is left in one row (7 = 2 × 3 + 1).*

Figure 2. *The largest prime known in 1979, having 6,987 digits. In 1986 a much larger prime was discovered, having more than 65,000 digits. Since the number of primes is infinite, it is only a question of time before still a larger prime will be found.*

$$2^{23209} - 1$$

```
                                       4028741157789887781818733290715917677224385068916224200410299635786945952408874008676 39
861461466537103833299413586592359075505942560215384203202392505282949656965468129867024629367955981
392588621343055240731175140117108509575771557641345266980711540248665702916557436111065215526167924 3
598639770340862427532627451407760338106482412910566360721371747330717161563227696645437427679839693 18
459384059938379878060275572931297882085263728828111767972047853647539734797508580059680824143812666 539
681419972077582269402957142281112038455662197412621809876912104670385133146958044938963769305794546 6
410758066275109556414379225335014944696332706216486506828452694695378794886597731643600416303767015 6
288981621675415729125137439286714883498899856647713325489027165934178641826671600314908207705819130
286776273368730270198911017821994203875686749027085635603356654222499933446410042566253239062428308 1
846894289907423494999391385972975650914894079376223505239789947674724335448214234181856675079971966
290454555100106794831122873316939190687807078070609507534444543783013175183099338301868495249460351 4
377326291391463671204785509870160512816327881326056457208975094051001640878295466349732839019544150 2
031731812102334342144210092424063044916256832054226602502916389870929519931364601129148550918588500
543071567482692038979207121165184595857972650794067721766663304877552877530735512541831254657221163 6
885200104813266702629472920752149577319512193931311360717000667227690006706061913909910619732429675 1
291126435194704329698008416655286098747171661877631026731179928184982563457209458635156898000127000 7
699500229311699961571595176656828314289491943888238134356695974575348674606207374249015590114664737
390675524203141974659857345920454694440079875042682825674223619120679037828851992942537332753463401 0
438383561431588941953576121530243393174665084559433694970370380773730716496281597828610083382864120 6
436538659807846921750141914508632328601539864514384665868751711078980714897238167180581728616753062 5
075117472492458080059976221116418632001809075690481059645707924323186702496567189414787013866778869 7
233847547755075627804890874585585714039616084091564000015294275076757908010594810079123600611381602 7
294954023517657545229368410211569897436302803253236256834802174708129208380881582322497874260497465 1
803265653773566324633198814887977168788134020708650038333581093114016383997899403087553195715489167 7
944261644640762439843597473725485956721633030725124135610398269230767188443842039966471204574771734 9
853326279186496926624767028353656166351203952980949968134964402925376450259648067636179607575343175
999065451862857799363623964096315727896274973523874018031963137162294519353089671616839856140612 93
442337629280035700233794829119523189191631925458763802371999173989413535766750903889069376224029198 0
336551006243082475037374270454634055589954829529267653294431255919681866264708150515182120867007206 2
662244135785787663162882922000802109909859588086490459353019579199283600567621973722016336401432076 98
917344225895364703243055887715189252485255414616714688385376991056124632069470551245708892171197864 4
056703135770947386634729787650241736975543242575119541062976565017144454111153411865771642670096
665217784040079089425722044373575364470132023204490873743591198846888965706531536359656993498881668 1
656834670402770338901203538825863885273372725296802331175711226959450369433557823964169012365540952
244084485151548107189417781442530197357722865554948884829926768347366006630669243185055068674539944 5
520961390335666138155130317070769706655969456682106844476806505097864405738968952295270831628476
713602096813200521662297702228641383037720127445077295416522867251086259308924256949130546482529205 1
070670232638922508876510803630212114342714333976089930833036430073258557653589204400883287541311598
336344115299539189336865789177906496416864360423327742300216028240865056235601373753305378621159873 6
805557064344886640795852215801092988641923941455936928232399558235098270802153492653110351340097081 2
402989211058114698144382506102890169695315562317736449187133984510136109015630882603126705678523924 0
688854691321940858515831306005272082801995694035599824708223899549815719733824216351158321846675607
771810334138571606559041577651060598586410945242414351010571013653342029099514179370387940205512396 29
386981798285845890434316234615652170924747125041080955071851370843614062903749489753378840376893337
967019452014316205535398240557240745650789504233993776048710965080738857416484988531003380895803114
702386171393470918048735414818380392458071193268641579841938267935480100978606147490699272167835318 44
871948502499906369821124257836674239877657956870261400878958629772354829973768267233153741132379149 0
133380537171174143349785173683206999265385587000255356918684065909860721703394780414441674592801942 4
098692156435119983976508231816318389379221957628088043293326015802569692130149889856075048804362100 06
948539607383883986631320236873887726358278654446890404384228812146483253624666106598128595021876652
656304044747719532951373264109049549817756983623389046479884300530772600312667698144843391181435818 7
830997562426197639363770484733809493659989183378814319610238163619306208581346901169487332307324999 2
898163744039295062511669662036979106898949654947337504706557753990341231870234321605170515557947761
192933448713338636368080179510596537097317217740629067636115194380331872850516648735434607676631522
745733922691440568189717446200213793419920060024109246994299643355073772649900571748863080009435280 2
758066832319193263432819944316604346404680550290495282346830424043178905143114575017478057302915937
660889742311665775464699337722520589967661164812372064100701891617584659685066100701891617 ...
```

The first fact was proven already by Euclid, in the third century B.C., and is one of the very few facts about the primes we know for sure (his proof is given in the Appendix). The second fact has given rise to many speculations and has endowed the primes with a certain air of mystery. There have been numerous attempts to discover a "prime producing formula," some mathematical expression which would permit us to predict the next prime number after a given one (just as the formula 2^n predicts all powers of 2). So far, all these attempts have failed. The primes seem to be distributed irregularly among the integers, and except for the very first few integers, it is next to impossible to tell whether a number is prime by merely looking at it (i.e., without actually trying to divide it by smaller integers). We do know, however, that the *average* distribution of the primes decreases as one moves to higher numbers; that is, the primes become more and more sparse. Thus between 1 and 100 there are 25 primes; between 100 and 200 , 21 primes; between 200 and 300, 16 primes; and so on. (However, there are 17 primes between 400 and 500, showing again that only the average distribution is of importance.) One of the major breakthroughs in our understanding of the primes came in 1896 when Jacques Hadamard (1865–1963) and de la Vallée Poussin (1866–1962) confirmed a conjecture made almost a century earlier by Carl Friedrich Gauss (1777–1855) regarding the statistical distribution of the primes. Their theorem, known as the Prime Number Theorem, says that the density of the primes—the number of primes below a given integer N, divided by N—tends to the number $1/ln\ N$ (where $ln\ N$ is the natural logarithm of N), as N tends to infinity.

Many questions about the primes remain unanswered, among them some which look deceptively simple. For example, primes have a tendency to crowd in pairs of the form $p,\ p+2$: 3 and 5, 5 and 7, 11 and 13, 41 and 43, 101 and 103, and so on. One finds these pairs even among very high numbers: 29,669 and 29,671, 29,879, and 29,881. An unsolved question is whether these "twin primes" are finite or infinite in number. Most mathematicians believe that there are infinitely many twin primes—just like the primes themselves—but all attempts to prove this conjecture have thus far failed. (In 1982 a computer software company offered a prize of $25,000 for a proof of this conjecture.) The largest twin primes known in 1963 were

$$140,737,488,353,699$$

and

$$140,737,488,353,701$$

but recently a much larger pair has been discovered:

$$1,159,142,985 \times 2^{2304} \pm 1$$

(each of these numbers has 703 digits).[1] Of course, new *single*

---

[1] C. W. Trigg: *J. Recreational Mathematics*, 14 (1981–82), 204.

primes are being discovered each year; the largest one known as this book goes to print (January, 1986) is $2^{216,091} - 1$, a 65,050-digit number. It was discovered by David Slowinski of the Cray Research company in Chippewa Falls, Wisconsin, using a testing program for new supercomputers.

The Fascination of Infinite Series

> *Even as the finite encloses an infinite series*
> *And in the unlimited limits appear,*
> *So the soul of immensity dwells in minutia*
> *And in narrowest limits no limit in here.*
> *What joy to discern the minute in infinity!*
> *The vast to perceive in the small, what divinity!*
>
> — Jacques Bernoulli (1654–1705)

A *series* is obtained from a sequence by adding up its terms one by one. From a finite sequence $a_1, a_2, a_3, \ldots, a_n$ we obtain the finite series, or sum, $a_1 + a_2 + a_3 + \ldots + a_n$. But for an *infinite* sequence $a_1, a_2, a_3, \ldots, a_n, \ldots$, a problem arises: How should we compute its sum? We cannot, of course, actually add up all its infinitely many terms; but we can, instead, sum up a finite, but *ever increasing,* number of terms: $a_1, a_1 + a_2, a_1 + a_2 + a_3$, and so on. In this way we obtain a new sequence, the sequence of *partial sums* of the original sequence. For example, from the sequence $1, 1/2, 1/3, \ldots, 1/n, \ldots$ we get the sequence of partial sums $1, 1 + 1/2 = 1.5, 1 + 1/2 + 1/3 = 1.83333\ldots$, and so on. If this sequence of partial sums converges to a limit S, then we say that the infinite series $a_1 + a_2 + a_3 + \ldots$ *converges to the sum S.* For the sake of brevity, we also use the phrase, "the series has the (infinite) sum S."

Once again, a special notation is used to describe this situation:

$$a_1 + a_2 + a_3 + \cdots = S$$

or, alternatively,

$$\sum_{i=1}^{\infty} a_i = S$$

(The symbol $\sum$ is the Greek capital letter sigma and stands for "the sum of"; the subscript i is a variable whose value "runs" from 1 to infinity.) Note that these equations are really abbreviations for the statements:

$$\lim_{n \to \infty} (a_1 + a_2 + a_3 + \cdots + a_n) = S$$

or, alternatively,

$$\lim_{n \to \infty} \sum_{i=1}^{n} a_i = S$$

But how can we tell whether a given series *has* a limit, and if so, what that limit is? The answer to the first question is relatively easy (but requires some knowledge of calculus); for the second question there generally is no answer available.

We can easily tell when a series does *not* converge to a limit: this happens whenever its terms become larger and larger. It is hardly surprising that the series $1 + 2 + 3 + \ldots$ does not converge, since its partial sums keep growing beyond all bounds. A series that does not converge to a definite sum is said to *diverge.*

The surprise comes when we discover that there are series whose terms get smaller and smaller, and still they do not converge! The classic example of this is the *harmonic series,* obtained by adding the reciprocals of the natural numbers:[1]

$$1/1 + 1/2 + 1/3 + 1/4 + \cdots$$

Here the terms certainly become ever smaller, and yet the series refuses to converge. Slowly—very slowly—its sum will grow and exceed any finite value. Thus the requirement that the terms of a series get smaller and smaller as $n \to \infty$ is a *necessary,* but not a *sufficient,* condition for its convergence.

The strange behavior of the harmonic series has puzzled—and fascinated—mathematicians for generations. Its divergence was first proved by Nicolae Oresme (1323?–1382), a French scholar, nearly four centuries before the limit concept became fully understood. (His proof is the same one we use today and is given in the Appendix.) Yet if we set out to add the terms of the harmonic series, there is nothing in the behavior of the partial sums that will give us even the slightest hint that the series might diverge. Some figures will make this clear: If we sum up the first thousand terms of the series, the partial sum will be 7.485, to the nearest thousandth; for the frst million (10^6) terms, it will be 14.357; for the first billion (10^9) terms, approximately 21; for the first trillion (10^{12}) terms, approximately 28; and so on. Or we may

---

[1] The adjective "harmonic" comes from a certain connection between the members of this series and the intervals of the musical scale.

put the question in reverse: How many terms would we have to sum up to make the partial sums exceed a given number N? For $N = 5$, we would have to add up 83 terms;[2] for $N = 10$, 12,367 terms; for $N = 20$, approximately three hundred million terms. But suppose we decide to be really ambitious and let $N = 100$. A simple estimation formula, taken from calculus, shows that we would have to add up approximately 10^{43} terms—that is, 1 followed by 43 zeros! Now this is certainly a vey big number, but just how big it is can be grasped from the following calculation: Suppose we give the task of summing up our terms to a computer capable of adding one million terms per second. It will take the computer some 10^{37} seconds to sum up our 10^{43} terms. Now the age of the universe is presently estimated to be "only" about 10^{17} seconds; thus, it will take our machine many "universe lifetimes" to do the job! We may put the situation in yet another way. Let us imagine that we try to write down the series, term by term, on a long paper ribbon until its sum exceeds 100. Assuming that each term occupies one centimeter along our ribbon (which is certainly an underestimation, since the terms require more and more digits), we would have to use up 10^{43} centimeters of ribbon, which is approximately 10^{25} light-years.[3] But the known size of the universe is estimated at only 10^{12} light-years, so that the entire universe will not be large enough to write down our series until its sum exceeds a modest 100. And yet we can be assured that if we could sum up the *entire* series—that is, all its infinitely many terms—the sum would indeed become infinite. Such is the nature of the infinite process that some series converge to their limits, while others which seem to converge simply refuse to do so.[4]

The peculiar behavior of the harmonic series has given rise to many interesting questions, some of which have not yet been answered. The peculiarity, as we have seen, comes from the fact that the terms get smaller and smaller, and yet the series diverges. Some terms, then, must be responsible for this divergence. The question therefore suggests itself: Which terms—or groups of

---

[2] It is easy—if somewhat tedious—to do this on a simple calculator. A programmable calculator, or better yet, a computer, will of course do the job much faster.

[3] A light-year is the distance that light, traveling at about 300,000 km/sec, covers in one year. It is approximately 9.4×10^{12}, or very nearly ten trillion, kilometers.

[4] See R. P. Boas, Jr. and J. W. Wrench, Jr., "Partial Sums of the Harmonic Series," *the American Mathematical Monthly,* 78 (1971), pp. 864–870.

terms—must be removed from the series to make it converge? Much research has been done on this question. It has been proved, for example, that if we remove from the series all the terms whose denominators are composite numbers (such as 1/4, 1/6, 1/8, 1/9, etc.), the resulting series will still diverge! This is rather remarkable, since the remaining terms are the reciprocals of the prime numbers (see p. 21), and the primes become increasingly sparse as we move to higher numbers. Thus one would intuitively expect that by depleting the harmonic series so that it includes only the reciprocals of the primes, the series would converge. But this is not so.

On the other hand, the series consisting of the reciprocals of all *twin* primes—pairs of primes of the form p and $p + 2$ (such as 3 and 5, 5 and 7, 11 and 13, and so on)—does indeed converge. But a word of warning must be added here: we do not know whether these twin primes are finite or infinite in number. Most mathematicians believe that their number is infinite (even though they are distributed even more sparsely among the natural numbers than the primes themselves); but until we know for sure— and this may not be very soon—we are not entirely justified in regarding this series as an infinite one.

Other attempts at removing terms from the harmonic series have been made. Let us mention one more: If we delete from the series all terms that have the digit 9 in their denominator (such as 9, 19, 92, 199, etc.), the resulting series is known to converge to a sum between 22.4 and 23.3.[5]

---

[5] See Frank Irwin, "A Curious Convergent Series," *the American Mathematical Monthly,* 23 (1916), pp. 149–152.

The Geometric Series   5

With the exception of the geometric series, there does not exist in all of mathematics a single infinite series whose sum has been determined rigorously.

— Niels Henrik Abel (1802–1829)

If the harmonic series is the most celebrated of all divergent series, the same distinction for convergent series goes, without reservation, to the geometric series. We have already met this series in connection with the runner's paradox. In a geometric *sequence,* or progression, we begin with an initial number a and obtain the subsequent terms by repeated multiplication by a constant number q: $a, aq, aq^2, \ldots, aq^n, \ldots$. The constant q is the *common ratio,* or quotient, of the progression. Sometimes our progression is terminated after a certain number of terms, in which case, of course, we omit the final dots. Such finite geometric progressions appear quite frequently in various situations. Perhaps the most well known is compound interest: If one deposits, say, $100 in a savings account that pays 5% annual interest, then at the end of each year the amount of money will increase by a factor of 1.05, yielding the sequence $100.00, 105.00, 110.25, 115.76, 121.55, and so on (all figures are rounded to the nearest cent).[1] On paper, at least, the growth is impressive; alas, inflation will soon dampen whatever excitement one might have derived from this growth!

The rate of growth of a geometric progression can be quite astounding. There is a legend about the Shah of ancient Persia, who was so impressed by the newly invented game of chess that he asked to see the inventor and bestow on him the riches of

---

[1] One can easily obtain these figures on a calculator by using the "constant" feature, which enables one to do repeated multiplication by merely pressing the "equals" key as many times as needed.

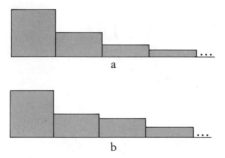

Figure 5.1. *A comparison of the geometric and harmonic series, each represented as the sum of the areas of rectangles of unit base. (a) The geometric series: each rectangle is one-half as high as its predecessor. The sum of the areas converges to the limit 2, i.e., twice the area of the initial square. (b) The harmonic series: the heights follow the sequence 1, 1/2, 1/3, 1/4, 1/5, ... This series diverges, so that the total area can be made as large as we please by adding more and more rectangles.*

the royal palace. When summoned to the king, the inventor, a poor peasant who was, however, well versed in mathematics, merely requested that one grain of wheat be placed on the first square of the chess board, two grains on the second square, four grains on the third, and so on until the entire board would be covered. The Shah, stunned by the modesty of this request, immediately ordered a sack of grain to be brought in, and his servants patiently began to place the grains on the board. To their utter astonishment, they soon discovered that neither the sack, nor the entire amount of grain in the kingdom, would suffice to fulfill the task, for the 64th term of the progression 1, 2, 4, 8, 16, 32, . . . is quite a large number: 9,223,372, 036,854,775,808. If we tried to place that many grains in a straight line, the line—assuming that each grain is one millimeter in diameter—would be some two light-years long![2]

Many people feel some sort of fascination with the way a geometric progression grows. Somehow they feel that the "natural" way a progression should grow is linearly, that is, by equal increments, as in the progression of natural numbers. Such a sequence is known as an *arithmetic progression* (the adjectives "arithmetic" and "geometric" are used simply to distinguish between the two types). Arithmetic progressions are also quite common: in a flight of stairs the height grows linearly, the street lights along a highway are separated by equal distances, and the ticks of a clock occur at equal time intervals. The difference between the geometric and arithmetic progressions can be seen from our savings account ex-

---

[2] The word "chess," incidentally, comes from "shah," giving some credence to the legend.

30

Figure 5.2. *Repeated reflections of a boy in a mirror. Each image is exactly one-half as high as its predecessor, thus following a geometric progression with q = 1/2. Photograph by Charles Eames from the exhibition* Mathematica *designed by Charles and Ray Eames for IBM, reprinted with permission.*

ample: had the bank paid simple instead of compound interest, the amount of money would have been $100, 105, 110, 115, 120, and so on. We see that from the third term on, the geometric progression begins to outgrow the arithmetic.

But of far greater interest to us here are *infinite* geometric progressions. These are just as common as their finite counterparts. A good example is radioactive decay: From an initial quantity of radioactive substance only half will remain after a certain period of time; after another equal lapse of time, one-fourth will remain, and so on. (Of course, the decaying substance will transform into other elements such as lead, and in the process radiation energy will be emitted.) Theoretically, this process will go on forever; in reality, it will come to an end when the last atom has been transformed. The period of time it takes for any initial quantity to decay to one-half of itself is its "half-life," and it greatly varies from one substance to another. The half-life of the common isotope of uranium (^{238}U) is about five billion years; that of ordinary radium (^{226}Ra) is 1,600 years. Another isotope of radium, ^{230}Ra, has a half-life of one hour, while that of ^{220}Ra is only 23 milliseconds. It is for this reason that many of the unstable elements are not found in natural minerals; whatever initial quantity there was when the earth was created has long since been transformed

31

into the more stable elements. The half-life is characteristic of the radioactive material in question and does not depend on the initial amount of that material: it takes the same 1,600 years to transform one gram of ^{226}Ra into half a gram as it takes to transform one ton into half a ton. It is this fact which is the basis of the famous carbon-14 test used in dating archeological finds.

We now come back to infinite geometric *series*. We have already seen that a necessary condition for any infinite series to converge is that the terms should become smaller and smaller as we go farther out. For the geometric series this means that the common ratio q must be a proper fraction, i.e., a number between -1 and 1. It turns out that, in this case, this is also a sufficient condition: whenever q is between -1 and 1, the series will converge. Moreover, the geometric series is one of the very few series for which we can exactly predict the limit; some readers may remember the formula from their high school algebra course:

$$a + aq + aq^2 + \cdots = \frac{a}{1-q}, \qquad \textit{if and only if} -1 < q < 1.^3$$

A proof of this important formula is found in the Appendix. For $a = 1/2$ and $q = 1/2$, it gives the limit for the series appearing in the runner's paradox: $1/2 + 1/4 + 1/8 + \cdots = (1/2)/(1 - 1/2) = 1$. A repeating decimal is another case in point: the decimal $0.999\ldots$ (often written as $0.\overline{9}$) is merely an abbreviation for the geometric series $9/10 + 9/100 + 9/1,000 + \cdots$, for which $a = 9/10$ and $q = 1/10$. This series converges to the limit $(9/10)/(1 - 1/10) = 1$, and we are thus perfectly correct in saying that $0.999\ldots$ is *equal* to 1—not just approximately but exactly. Many people find it hard to accept this simple fact, and one can often hear a heated discussion as to its validity.

The closer the common ratio q gets to 1 or -1, the slower the series converges to the limit predicted by the above formula; that is, more terms will be needed to get within a specified distance from this limit. For $a = 1$ and $q = 1$ we get the series $1 + 1 + 1 + \ldots$, which obviously diverges. For $a = 1$ and $q = -1$, however, an interesting situation arises: the series become $1 - 1 + 1 - 1 + - \ldots$, which does not tend to any definite limit, and yet our formula predicts that we should get the limit $1/2$. Something must have gone wrong! And sure enough—we have used our formula illegally, for it is valid only for values of q *between* -1 and 1, not *equal* to -1 or 1. Yet this case, we feel,

---

³ The phrase "if and only if" is synonymous to "necessary and sufficient."

is somehow different from the case where q is actually greater than 1 in absolute value (and for which the partial sums grow beyond all bounds); here the partial sums merely oscillate between 1 and 0:

$$1=1, \ 1-1=0, \ 1-1+1=1, \ 1-1+1-1=0, \ \text{and so on}$$

This seemingly bizarre behavior was the source of much controversy in the years following the invention of the calculus. Leibniz himself argued that since the sum may be 0 or 1 with equal probability, its "true" value should be their mean, which is 1/2—in agreement with our formula![4] Such a brazen, careless reasoning indeed seems incredible to us today, but in Leibniz's time the limit concept was far from fully understood, and infinite series were treated in a purely manipulative manner, without regard to the question of their convergence. Today we know that this question is absolutely crucial: a divergent series simply *does not* have a sum, and any attempt to assign to it a specific numerical value is doomed to fail. One of the first mathematicians to realize this fact was the Norwegian Niels Henrik Abel (1802–1829), who in 1828 wrote:

> *The divergent series are the invention of the devil, and it is a shame to base on them any demonstration whatsoever. By using them, one may draw any conclusion he pleases and that is why these series have produced so many fallacies and so many paradoxes. . . .*[5]

---

[4] One can also arrive at other values for the sum of this series: If we write our series as $1 + (-1 + 1) + (-1 + 1) + \ldots$, its sum is seen to be 1; if we write it as $(1 - 1) + (1 - 1) + \ldots$, its sum is 0. But then again if we call our sum S, we may write $S = 1 - (1 - 1 + 1 - 1 + \ldots) = 1 - S$, from which we conclude that $S = 1/2$. All these "conclusions," of course, are false, because the series diverges.

[5] Abel was wrong, however, in implying that divergent series are useless; in fact, they have found many applications and are an important tool in analysis. See Godfrey Harold Hardy, *Divergent Series,* Oxford University Press, 1949.

6 More about Infinite Series

In the use of this method (of infinites) the pupil must be awake and thinking, for when the infinite is employed in an argument by the unskillful, the conclusion is often most absurd.

— Elisha S. Loomis (1852–1940)

Infinite geometric sequences and series arise not only in pure mathematics but also in geometry, physics, and engineering, and at least one contemporary artist, Maurits C. Escher, has based on them many of his works. We shall examine some of these in the following chapters. Meanwhile, let us take a brief look at some other series, several of which mark important milestones in the history of mathematics. We have seen that the harmonic series $1 + \frac{1}{2} + \frac{1}{3} + \frac{1}{4} + \ldots$ diverges. But the corresponding series with the *squares* of the natural numbers has baffled mathematicians for many years; among them were several of the Bernoulli brothers, who all failed to find its sum, although it had been known for some time that the series converges.[1] It was the great Swiss mathematician Leonhard Euler (1707–1783) who finally solved the mystery. His result, discovered in 1736, was as unexpected as it was elegant:

$$1/1^2 + 1/2^2 + 1/3^2 + 1/4^2 + \ldots = \pi^2/6$$

The proof is far from elementary, but the interesting point is that Euler used methods which today would not pass the critical

---

[1] The Bernoulli family was to mathematics what the Bach family was to music. For more than two centuries they dominated the mathematical scene in Europe, contributing to almost every branch of mathematics then known. At least twelve members of the family achieved mathematical prominence, beginning with Nicolaus Bernoulli (1623–1708). Most came from Basel, Switzerland, where they lived and worked. It was Jacques Bernoulli (1654–1705) who first proved that the series $1/1^2 + 1/2^2 + 1/3^2 + \ldots$ converges.

eye of any self-respecting mathematician. The surprise in Euler's discovery is the unexpected appearance of the number π in the limit of a series involving only the natural numbers. To this day, Euler's series is regarded as one of the most beautiful results in mathematical analysis.

Using similar methods, Euler was able to find the sum of many other infinite series involving the natural numbers. These results all appeared in 1748 in his monumental work *Introductio in analysin infinitorum*, which was entirely devoted to infinite processes and which is regarded as the foundation of modern analysis. Among his many results is the summation of the series $1/1^k + 1/2^k + 1/3^k + 1/4^k + \ldots$ for all even values of k from 2 to 26; for this last value he found the sum to be:

$$\frac{2^{24} \cdot 76{,}977{,}927 \cdot \pi^{26}}{1 \cdot 2 \cdot 3 \cdot \ldots \cdot 27}$$

The same series for *odd* values of k, however, is much more difficult to handle, and until very recently, the exact nature of the sum for the case $k = 3$ was not known.[2]

Euler was not the first to discover an infinite series that has some connection to π. We have already met Gregory's series, discovered in 1671:

$$1 - 1/3 + 1/5 - 1/7 + - \cdots = \pi/4$$

This series was one of the first results of the newly invented calculus; however, if we try to calculate the value of π from it, a disappointment will be waiting for us: the series converges so slowly that it requires 628 terms to approximate π to just two decimal places (that is, to 3.14)! It is much more practical to use some other formula, such as:

$$1/1^4 + 1/2^4 + 1/3^4 + 1/4^4 + \cdots = \pi^4/90$$

---

[2] It is known, however, that the general series $1/1^k + 1/2^k + 1/3^k + \cdots$ converges for all values of k greater than 1 and diverges for $k \leqslant 1$. (The case $k = 1$ is the harmonic series.) This theorem is proved in calculus. For the case $k = 3$, see Alfred Van der Poorten, "A Proof that Euler Missed . . . — Apéry's Proof of the Irrationality of $\zeta(3)$," *the Mathematical Intelligencer*, 1 (1979), pp. 195–203. Apéry in 1978 proved that the sum of this series is approximately 1.202. The series $\sum_{n=1}^{\infty} 1/n^k$, when regarded as a function of the exponent k (where k can assume complex values) is known as the zeta function and denoted by $\zeta(k)$. It shows up in various branches of mathematics and is the subject of active research.

which is Euler's series for $k = 4$ and which converges very fast. These examples highlight the difference between the theoretical notion of convergence and the practical question of the *rate* of convergence, a question which has become especially important since the advent of the computer.

Associated with infinite series are some of the most remarkable paradoxes of infinity, paradoxes which puzzled the seventeenth- and eighteenth-century mathematicians in much the same way as the paradoxes of motion had puzzled the Greek philosophers two thousand years earlier. For example, elementary arithmetic teaches us that in any *finite* sum we may rearrange the order of the terms without affecting the value of the sum. Thus $1 + 2 + 3$ is the same as $2 + 1 + 3$, which is the same as $3 + 2 + 1$. (In technical terms, we have used the associative and commutative laws in going from the first expression to the others.) The last sum is the first written in reverse; we often use such a reversing procedure in accounting, when checking the sum of a long list of numbers against any possible error. But can we also do this with *infinite* sums? Here, of course, we cannot write the sum in reverse, beginning with the last term and ending with the first, because there *is* no last term. But we can still rearrange our sum by, say, advancing some terms to a more forward position and displacing others to a later position in the series. The question is: Does such a rearrangement affect the sum? The unexpected answer is *yes:* rearranging the terms of an infinite series may, under certain conditions, affect the limit to which the series converges, and it may even change a convergent series into a divergent one! A classic example of this is furnished by the harmonic series with alternating signs, which is known to converge to the natural logarithm of 2 (written *ln* 2):

$$1/1 - 1/2 + 1/3 - 1/4 + 1/5 - 1/6 + 1/7 - 1/8 + - \cdots$$
$$= ln\ 2$$

Let us multiply both sides of this equation by $1/2$:

$$1/2 \qquad -1/4 \qquad +1/6 \qquad -1/8 + - \cdots$$
$$= \tfrac{1}{2}\,ln\ 2$$

We now add the two series term by term, adding the terms vertically:

$$1/1 \qquad +1/3 - 1/2 + 1/5 \qquad +1/7 - 1/4 + 1/9 + 1/11$$
$$-1/6 + - \cdots = \tfrac{3}{2}\,ln\ 2$$

But the last series is simply a rearrangement of the original one, and yet its sum is $3/2$ of the sum of the original series, leading to the "conclusion" that $1 = 3/2$! *"It is easy to imagine,"* said Richard Courant in his treatise on the calculus, *"the effect that the discovery of this apparent paradox must have had on the mathematicians of the eighteenth century, who were accustomed to operate with infinite series without regard to their convergence."*

The cause of this paradoxical behavior lies in the fact that the series $1 - 1/2 + 1/3 - 1/4 + - \ldots$ converges solely because the terms have alternating signs and therefore partially "compensate" each other. But had we taken the terms in their absolute value (i.e., all with positive signs), we would have gotten the harmonic series, which diverges. Here, then, is a fundamental distinction between two classes of convergent series: those series which converge irrespective of the signs of their terms (called *absolutely* convergent series), and those whose convergence is merely due to the alternating signs of their terms (*conditionally* convergent series). It is the former class which represents the stronger type of convergence, for here the convergence takes place because the terms themselves approach zero rapidly enough. It is proved in calculus that only in absolutely convergent series can a rearrangement of terms be done without affecting the sum. Here, then, we get the first inkling that the ordinary rules of arithmetic, always valid for finite calculations, may break down when the infinite is involved.

The series we have discussed here all belong to a very large and important class of series known as *power series*. Such series have the general form $a_0 + a_1 x + a_2 x^2 + \cdots$, or briefly $\sum_{n=1}^{\infty} a_n x^n$, where the a_n's are the coefficients, or constants, while x is a variable whose value may determine whether the series converges nor not.[3] For example, the geometric series $1 + x + x^2 + \cdots$ is a power series for which all the a_n's are equal to 1, while x must have any value between -1 and 1 if the series is to converge. Similarly, the harmonic series is obtained from the

---

[3] Sometimes such a series is referred to as an "infinite polynomial," indicating the extension of the polynomials studied in elementary algebra to infinitely many terms. (A *polynomial* is an expression of the form $a_0 + a_1 x + a_2 x^2 + \cdots + a_n x^n$; the a_i's are the *coefficients*, and the number n, the highest power of x present, is the *degree* of the polynomial. For example, the polynomial $3 - 2x + 7x^2$ is of degree 2, and its coefficients are $a_0 = 3$, $a_1 = -2$, and $a_2 = 7$.) The use of the term "polynomial" for an infinite power series is, however, not entirely justified, since, as we have seen, such series may not obey all the rules of arithmetic that finite polynomials obey.

power series $x + x^2/2 + x^3/3 + \cdots$ by letting $x = 1$. To gain a complete insight into these power series—and especially into the question of their convergence—we must go beyond the real number system into the domain of *complex numbers*. These are numbers of the form $x + iy$, where x and y are ordinary real numbers and i is the celebrated "imaginary unit," the square root of -1. Although the technical details of this extension are beyond our scope here, we can at least outline them briefly. In order to tell whether a given power series converges, we must look at the series in the complex domain, that is, regard all its terms (including the coefficients) as complex numbers.[4] Then every series is associated with a circle, called the *circle of convergence*, inside which the series will always converge, while outside it will diverge. What happens *on* the circle of convergence cannot, in general, be predicted: a series may converge for some points on the circle and diverge for the others, or it may converge for all points on that circle, or again it may converge for none. For example, the geometric series $1 + z + z^2 + \cdots$ (where z now stands for the complex number $x + iy$) has as its circle of convergence the unit circle (the circle with center at O and radius 1), and it *diverges* for all points on this circle. (We may check this for the points 1, -1, i, and $-i$, all of which lie on the circle; see Fig. 6.1.) On the other hand, the series $z + z^2/2 + z^3/3 + \cdots$, which has the same circle of convergence, diverges for $z = 1$ (giving the harmonic series) but converges fot $z = -1$, having $-ln\ 2$ as its sum. There are also power series that converge for *all* values of z, in which case the circle of convergence has an infinite radius; others converge for *no* value of z except $z = 0$, and the circle of convergence shrinks to a single point.

With the discovery of these facts by the French mathematician Augustine Louis Cauchy (1789–1857) in 1831, a full understanding of infinite series was finally achieved. From then on, power series became an indispensable tool in almost every branch of mathematics, leading to their extension to more general types of series where the terms are not powers of z (i.e., z^n) but more complicated functions of z, notably trigonometric functions.[5] This culminated in the discovery that a function itself—almost any function—can be represented as an infinite series of some sort, a discovery that has found numerous applications in physics and played a crucial role in the development of quantum theory. At the same

---

[4] A *real* number is a special case of a complex number. The number 5, for example, is the complex number $5 + 0i$.

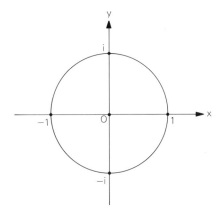

Figure 6.1. *The circle of convergence for the series* $1 + z + z^2 + z^3 + \cdots$, *where* $z = x + iy$. *The symbol* i *stands for the "imaginary unit,"* $\sqrt{-1}$.

time the theoretical foundations of analysis were shaped and re-fined, and the rules of operation with infinite processes laid down, until the last shred of doubt was removed and every paradox resolved to the satisfaction of even the most rigor-minded mathematician.

It seemed, then, towards the end of the nineteenth century, that the long struggle to grope with and understand the concept of infinity had come to its successful conclusion, and that the concept was finally standing on firm foundations. It was then that a young and relatively unknown mathematician came on the scene and shook these foundations to the ground. His name was Georg Cantor.

---

[5] A *function* in mathematics is a rule that assigns to every permissible value of one variable (called the *independent variable* and often denoted by x or z) a single value of a second variable (the *dependent variable*, often denoted by y). For example, the equation $y = x^2$ assigns to every value of x its own square, and hence is a function; likewise, the equation $y = 1/x$ assigns to every non-zero value of x its reciprocal. Functions of a more general nature will be considered in Chapter 11.

7 Interlude: An Excursion into the Number Concept

Number rules the universe.

— Motto of the Pythagorean School

God made the integers; all the rest is the work of man.

— Leopold Kronecker (1823–1891)

To appreciate Cantor's revolutionary ideas about the infinite, we must first make a brief excursion into the history of the number concept. The simplest type of numbers are, of course, the counting numbers 1, 2, 3, ... Mathematicians prefer to call them the *natural numbers,* or again the *positive integers.* Simple though they are, these numbers have been the subject of research and speculation since the dawn of recorded history, and many civilizations have assigned to them various mystical properties. An important branch of modern mathematics, number theory, deals exclusively with the natural numbers, and some of the most fundamental questions about them—for example, questions relating to the prime numbers—are without answer to this day. But without reservation, the single most important property of the natural numbers is this—*there are infinitely many of them.* The fact that there is no last counting number seems so obvious to us that we hardly bother to reflect upon its consequences. The entire system of calculations with numbers—our familiar rules of arithmetic—would have collapsed like a house of cards if there were a last number beyond which nothing else existed. Suppose for a moment that such a number did exist, say 1,000. Then not only would we have to ignore anything greater than 1,000, but all calculations that lead to numbers in excess of 1,000, such as 999 + 2 or 500 + 600, would become "illegal." Our ordinary art of computing, in other words, would have to be abandoned. Fortunately, this is not the case. We will accept the infinitude of the counting numbers as an *axiom,* a statement whose validity can be taken for granted.

Stated more formally, the axiom says: *Every natural number n has a successor, n + 1.*

If we add to the set of natural numbers the notion of direction, we get the set of *integers,* or *whole numbers.* These are made up of the natural numbers, their negatives, and the number zero. The integers do not differ in any essential way from the natural numbers except that they stretch indefinitely in both directions. Thus every negative number can be thought of as a "mirror image" of the corresponding positive number, a fact that is graphically illustrated on the familiar number line (Fig. 7.1).[1]

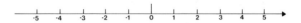

Figure 7.1. *The number line.*

Next in complexity are the ordinary fractions, or *rational numbers;* these are numbers of the form $\frac{a}{b}$ (also written a/b or $a:b$), where a and b are integers. The denominator b, of course, must never be zero, for division by zero is an illegal operation. Examples of rational numbers are $2/3$, $3/2$, $-17/9$, and also 5, because 5 can be written as $5/1$. Thus, the set of integers is a *subset* of the set of rational numbers (just as the natural numbers are a subset of the integers). Fractions were known to man almost as soon as he learned how to operate with counting numbers, for any measurement which does not exactly result in a whole number leads to the use of fractions. The Greeks, in particular, had a very high regard for fractions. They believed—and this was the essence of the Pythagorean teaching—that everything in nature should be expressible in terms of ratios of whole numbers. This philosophy, in most likelihood, had its origin in music. As Pythagoras himself had discovered, the common musical intervals produced by a vibrating string correspond to simple numerical ratios

---

[1] Simple though they are, the negative numbers were slow in being accepted as *bona fide* numbers into mathematics. The Greeks barred them altogether or treated them as "absurd" quantities, in much the same way as the "imaginary number" i (the square root of -1) was looked upon later. The first mention of negative numbers seems to be in the works of Brahamagupta, a Hindu mathematician who lived in the seventh century A.D. It was not until the seventeenth century that negative numbers were fully accepted into our number system. For a fuller discussion of their history, see the book *History of Mathematics* by David Eugene Smith, Dover Publications, New York, 1958, Vol. II, pp. 257–260.

of string lengths. To produce two notes one octave apart, we let the string vibrate first at its full length, then at half its length; thus an octave corresponds to the ratio 2:1. Likewise, the interval of a fifth corresponds to the ratio 3:2, a fourth to 4:3, and so on. In fact, the more pleasing the interval sounds—the more "consonant" it is—the simpler the fraction that represents it. Dissonant intervals have more complicated ratios, such as 9:8 for the second and 16:15 for the half-tone. Since in the Greek world music ranked equally in importance with mathematics and philosophy, the Greek scholars saw in these facts a sign that the entire universe is constructed according to the laws of musical harmony, that is, from fractions. Rational numbers thus dominated the Greek view of the world, just as rational thinking dominated their philosophy. (Indeed, the Greek word for rational is *logos,* from which our modern word "logic" comes.)

Just as there are infinitely many integers, there are infinitely many fractions. But there is one important difference: while the integers have big gaps between them, gaps of one unit each, the rational numbers are *dense,* meaning that between any two fractions, no matter how close they are, we can always find another fraction. For example, between 1/1,001 and 1/1,000 (which are certainly very close, their difference being about one-millionth) we can put the fraction 2/2,001. (Indeed, the decimal values of

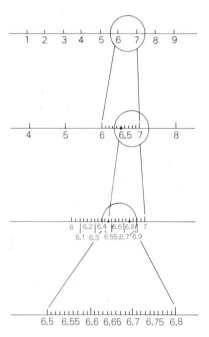

Figure 7.2. *The density of the rational numbers: between any two rationals, no matter how close, there are infinitely many other rationals.*

these fractions are 0.000999, 0.001, and 0.0009995, respectively.) We may now repeat the process and cramp another fraction anywhere between 2/2,001 and 1/1,000 (3/3,001 would be an example), and so on *ad infinitum* (Fig. 7.2). Not only, then, can we squeeze a new fraction between any two given fractions, we can in fact squeeze *infinitely many* fractions between them. But this is merely another way of saying that the process of division can be repeated indefinitely. Unlike the situation in the material world, where we will ultimately reach the atomic or subatomic level (the so-called "elementary particles"), there is no "mathematical atom," no smallest unit that cannot be divided into smaller units yet.

It seems, then, that the rational numbers form a vast, densely populated set of numbers, leaving no gaps between its members. This in turn would mean that the entire number line is completely populated with rational numbers, or "rational points." It was therefore one of the most momentous events in the history of mathematics when it was discovered that this is not so. In spite of their density, the rational numbers leave "holes" in the number line— that is, points that do not correspond to any rational numbers!

8 The Discovery of Irrational Numbers

The attempt to apply rational arithmetic to a problem in geometry resulted in the first crisis in the history of mathematics. The two relatively simple problems—the determination of the diagonal of a square and that of the circumference of a circle—revealed the existence of new mathematical beings for which no place could be found within the rational domain.

— Tobias Dantzig (1884–1956)

The discovery of these "holes" is attributed to Pythagoras, founder of the celebrated Greek school of mathematics and philosophy in the sixth century B.C. The life of Pythagoras is shrouded in mystery, and the little we know about him is more legend than fact. This is partially due to an absence of documents from his time, but also because the Pythagoreans formed a secret society, an order devoted to mysticism, whose members agreed upon strict codes of communal life. There is some doubt whether many of the contributions attributed to Pythagoras were indeed his own, but there is no question that his teaching has had an enormous influence on the subsequent history of mathematics, an influence which lasted for more than two thousand years. His name, of course, is immortally associated with the theorem relating the hypotenuse of a right triangle to its two sides, even though there is strong evidence that the theorem had already been known to the Babylonians and the Chinese at least a thousand years before him. The theorem says that in any right triangle, the square of the hypotenuse is equal to the sum of the squares of the two sides: $c^2 = a^2 + b^2$ (Fig. 8.1). The Pythagorean Theorem is probably the most well know, and certainly the most widely used theorem in all of mathematics, and it appears, directly or in disguise, in almost every branch of it.

Now among all right triangles there is one which is of special importance to our discussion: the right triangle which at the same time is also isosceles; that is, for which $a = b$. Since we are free to choose our unit of length at will, we may assign to each side a unit length ($a = b = 1$). The Pythagorean Theorem then says that $c^2 = 1^2 + 1^2 = 2$, and the hypotenuse c therefore has a length equal to the square root of 2, written $\sqrt{2}$. With the square root key on any calculator we can find its approximate value: 1.41421.

There exist various methods, or *algorithms*,[1] to find the approximate value of $\sqrt{2}$ to any desired accuracy, even though one can never find its "exact" value, since this would require an infinite number of digits (just as with π). But this fact does not prevent us from being able to locate the point corresponding to $\sqrt{2}$ on the number line, and locate it *exactly*. We choose as our unit of length the segment from 0 to 1 (Fig. 8.2). At the end point of this segment we erect a perpendicular, also of unit length. We now place a compass at 0, open it until its other arm coincides with the end point of the perpendicular, and swing an arc. The point where this arc intercepts the number line is the desired point, since its distance from 0 is equal to $\sqrt{2}$. Note that the entire construction uses only two instruments—a straightedge and a compass—in agreement with the Greek tradition which required that all geometric constructions be done with just these two tools. Our construction also shows that the desired point lies somewhere between the points 1 and 2 on the number line.

Figure 8.1. *The Pythagorean Theorem: In any right triangle, $c^2 = a^2 + b^2$ where a and b are the sides and c is the hypotenuse.*

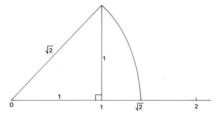

Figure 8.2. *Geometric construction of the irrational number $\sqrt{2}$, using only a straightedge and compass.*

But what kind of a number *is* $\sqrt{2}$? The Greeks, of course, assumed that it is a rational number, since this was the only kind of number they knew. But one day some unknown member of the Pythagorean school—perhaps Pythagoras himself—made the startling discovery that $\sqrt{2}$ is not commensurable with the unit—that is, the two numbers don't have a common measure. Now this is the same as saying that $\sqrt{2}$ cannot be written as a ratio of two integers; for if it could, then from the equation $\sqrt{2} =$

----

[1] An *algorithm* is a "recipe" consisting of a finite number of steps which, when followed, will lead to the solution of a mathematical problem. An algorithm for $\sqrt{2}$ would enable us to find its decimal expansion to any given (finite) number of decimal places. The word "algorithm" comes from Al Khowarizmi, an Arab mathematician living in the ninth century who was mainly responsible for bringing the Hindu system of numeration into widespread use in Europe.

a/b we would conclude that b times $\sqrt{2}$ is exactly equal to a units, making $\sqrt{2}$ and 1 commensurable. So here was something which clearly was a number (since it represented length), and yet which could be written neither as an integer nor as a ratio of integers. Thus the world first learned of the existence of irrational numbers.

The circumstances under which this discovery was made are shrouded in total mystery, and we do not even know what proof was used in the process. Today there are at least three different proofs of the irrationality of $\sqrt{2}$, one of which is given in the Appendix. By all likelihood, the Greeks used a geometric rather than algebraic proof, as it was the study of geometry that was their main interest (and besides, as we have seen, they did not possess the algebraic language). In any event, the discovery left them bewildered, for here was a geometric magnitude that defied their belief in the supremacy of rational numbers. So utter was their amazement that for a while they refused to regard $\sqrt{2}$ as a number at all, so that in effect the diagonal of a square had a numberless length! Legend has it that the Pythagoreans, fearing that the discovery might have adverse effects on the populace, vowed to keep it a closely guarded secret. But one of them, Hippasus, did breach out the news; whether his motives were purely academic or political we shall never know, for his fellows threw him overboard from the ship they were sailing, and his body rests to this day at the bottom of the Mediterranean.

There is yet another reason for introducing irrational numbers into mathematics. We know from elementary arithmetic that if we add, subtract, multiply, or divide two or more fractions, the result is again a fraction; for example, $1/1 + 1/4 + 1/9$ is $49/36$. Mathematicians call this property the *closure* of rational numbers under the operations of addition, subtraction, multiplication, and division. But this rule, like others we have already met, may break down when we attempt to extend it to an infinite sum or product. We have already seen that the infinite series $1/1 + 1/4 + 1/9 + 1/16 + \cdots$ converges to $\pi^2/6$ and the infinite product $(2 \cdot 2 \cdot 4 \cdot 4 \cdot 6 \cdot 6 \cdots)/(1 \cdot 3 \cdot 3 \cdot 5 \cdot 5 \cdot 7 \cdots)$ to $\pi/2$, both of which are irrational numbers and cannot therefore be written as fractions. Thus, the rational numbers are closed under the four basic arithmetic operations only so long as we apply these operations a finite number of times. When we apply them infinitely many times, the result may transcend the realm of the rationals.

Today the existence of irrational numbers disturbs no one anymore. In fact, not only the square root of 2 but also the square roots of all prime numbers ($\sqrt{3}$, $\sqrt{5}$, $\sqrt{7}$, $\sqrt{11}$, etc.) and of most composite numbers ($\sqrt{6}$, $\sqrt{8}$, etc.) are irrational, as are the numbers π and e and combinations of them. A completely satisfactory theory of irrationals was not given until 1872, when Richard Dedekind (1831–1916) published his celebrated essay *Continuity and Irrational Numbers.* The technical details need not concern us here; what matters is that the rational numbers, numerous as they are, are neither closed under the four basic arithmetic operations nor sufficient enough to cover the entire number line. They are *dense,* but they do not form a *continuum.* They leave many holes—infinitely many, in fact—and these holes are filled by the irrational numbers.

If we now unite the rational numbers with the irrationals, we get the larger set of *real* numbers. The adjective "real" is not meant to indicate any "real" character of these numbers; they are neither more nor less real than, say, the imaginary numbers or any other system of symbols used in mathematics. This is but one of the ironies of our spoken language that many a common word assumes a totally different meaning when taken up by the professionals. In any event, the real numbers may be described as all those numbers which can be written as decimals, such as 0.5, 0.121212. . ., −2.513, and so on. These decimals fall into three categories: terminating, such as 0.5; non-terminating and repeating, such as 0.1212. . . (usually written as $0.\overline{12}$);[2] and non-terminating, non-repeating, where the digits never recur in exactly the same order (an example would be 0.1010010001. . .). It is an elementary fact of arithmetic that decimals of the first two types always represent rational numbers (in our examples, 0.5 represents the ratio 1/2 and $0.\overline{12}$ represents 4/33)[3] while decimals of the third type represent irrational numbers. However, since we can write numbers and calculate with them only up to a finite number of digits, we must conclude that from a *practical* point of view there is no real distinction between rational and irrational numbers. That is to say, we may truncate the decimal expansion of an irrational number after a certain number of digits and get

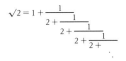

$$\sqrt{2} = 1 + \cfrac{1}{2 + \cfrac{1}{2 + \cfrac{1}{2 + \cfrac{1}{2 + \ddots}}}}$$

The square root of 2 written as an infinite continued fraction.

---

[2] Actually, a terminating decimal can be regarded as a repeating decimal with the digit zero repeating indefinitely after the last non-zero digit. For example, 0.5 = 0.5000. . . .

[3] This follows from the fact that a repeating decimal is really an infinite geometric series, and as we have seen on p. 32, such a series, if it converges, has the rational number $a/(1 - q)$ as its sum. To illustrate, the decimal $0.\overline{12}$ represents a geometric series with $a = 0.12$ and $q = 0.01$ (since the decimal can be written as $0.12 + 0.0012 + \cdots$); thus its sum is $0.12/(1 - 0.01) = 0.12/0.99 = 12/99 = 4/33$.

a *rational approximation* for our number, and this approximation can be made as good as we please by simply taking as many decimal digits as needed. (To illustrate, the decimals 1, 1.4, 1.41, 1.414 and 1.4142 are all rational approximations for $\sqrt{2}$, increasing progressively in their accuracy.) Thus, an engineer could hardly care less whether the length of an object is a rational or an irrational number, for even if the length is irrational (as in the diagonal of a square of unit sides), he could in any case measure it only with a limited degree of accuracy, due to the imperfections inherent in all measuring devices. These facts we must bear in mind as we now come to examine Cantor's revolutionary ideas about the infinite, for it is the theoretical rather than practical aspects of the number concept which are at the root of this revolution.

π, ϕ, and e —Three Celebrated Irrationals

There is a famous formula, perhaps the most compact and famous
of all formulas—developed by Euler from a discovery of de Moivre:
$e^{i\pi} + 1 = 0$ . . . It appeals equally to the mystic, the scientist, the
philosopher, the mathematician.

— Edward Kasner and James Newman

I believe that this geometric proportion served the Creator
as an idea when He introduced the continuous generation of similar
objects from similar objects.

— Johannes Kepler (1571–1630)

An irrational number is a number that cannot be written as the
ratio of two integers (whole numbers). The first number known
to be irrational was the square root of 2 ($\sqrt{2}$). The number π,
the ratio of the circumference of a circle to its diameter, was
known already to the Babylonians and Egyptians, and later Archimedes found its value to lie between $3^{10}\!/_{71}$ and $3\frac{1}{7}$. But it was
not until 1761 that the Swiss mathematician Johann Heinrich Lam-
bert (1728–1777) established the fact that π is irrational.

If we divide a line segment AB into two parts such that the
entire segment is to the long part as the long part is to the short
($AB/AC = AC/CB$, as in Fig. 1), then the point of division C
is said to divide AB in the "golden ratio." The numerical value
of this ratio is denoted by the Greek letter ϕ (phi). If we let
AB be the unit segment ($AB = 1$) and denote the length of AC
by x, then $\phi = 1/x = x/(1 - x)$. This leads to the quadratic
equation $x^2 + x - 1 = 0$, whose positive solution is $x =$
$(-1 + \sqrt{5})/2$, or approximately 0.61803. Therefore $\phi =$
$1/x = 1.61803. \ldots$ Like $\sqrt{2}$, ϕ is an irrational number whose
decimal expansion never ends and never repeats.

The golden ratio was known to the ancient Greeks and played
an important role in their architecture. Many artists believe that
of all rectangles, the one whose length-to-width ratio is ϕ has

Figure 1. *The golden section: C divides the segment AB so that the whole segment is to
the larger part as the large part is to the small.*

49

the "most pleasing" proportion—hence the prominent role this number has played in various theories of aesthetics. Surprisingly, it also shows up in the leaf arrangement of several plants. It has many interesting mathematical properties; for example, 1 divided by φ is equal to 1 less than φ (i.e., $1/\phi = \phi - 1$), as follows from the defining equation above. If we construct a "golden rectangle"—i.e., a rectangle with a length-to-width ratio equal to φ— then a second golden rectangle can be constructed whose length is the width of the original rectangle. This process can be repeated indefinitely, giving an infinite sequence of golden rectangles whose sizes diminish to zero (Fig. 2).

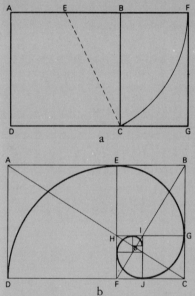

Figure 2. *Geometric construction of the number* φ: (*a*) *Start with a square of unit side, find the midpoint* E *of the side* AB, *and then, using* E *as center, swing an arc of length* EC *until it meets the extension of* AB *at* F. *The rectangle* AFGD *has a length-to-width ratio of* $(1 + \sqrt{5})/2$ *or* φ. (*b*) *By repeating this process indefinitely, we get a sequence of "golden rectangles" of ever smaller dimensions, until they converge at a point. The points* D, E, G, J *of these rectangles lie on a logarithmic spiral, a curve which we shall meet again in Chapter 11. Reprinted from H.E. Huntley,* The Divine Proportion: A Study in Mathematical Beauty, *Dover Publications, New York, 1970, with permission.*

The number *e* is the only one of the three celebrated irrationals that was not known to the ancients. Used as the basis for natural logarithms, it plays a central role in the calculus. It is the limit to which the expression $[1 + 1/n]^n$ tends as *n* approaches infinity. Its value, to five decimal places, is 2.71828; like all irrational numbers, it has a non-terminating and non-repeating decimal expansion, and therefore can never be written "exactly." The number e has an interesting and quite unexpected application in the financial world: If we could deposit \$1 in an imaginary savings account that pays 100% annual interest compounded *continuously*—i.e., at every instant, rather than annually or quarterly—we would expect the amount of money to grow beyond all bounds. Surprisingly, this is not so: after one year the sum will be equal to the number e, that is, to about \$2.72.

Even though $\sqrt{2}$, ϕ, π, and e are all irrationals, they belong to two fundamentally different classes of numbers: $\sqrt{2}$ and ϕ are *algebraic numbers*, that is, they are solutions of polynomial equations with integral coefficients. (Thus, $\sqrt{2}$ is the solution of the equation $x^2 - 2 = 0$ and ϕ is the solution of $x^2 - x - 1 = 0$). Numbers that are not solutions of such equations are called *transcendental*, and it is to this class that π and e belong. The transcendence of e can be proved relatively easily—the first proof was

Table 2. *Some infinite series and products.*

$$\phi = \sqrt{1 + \sqrt{1 + \sqrt{1 + \sqrt{1 + \cdots}}}}$$

Infinite Root
(discoverer unknown)

$$\phi = 1 + \cfrac{1}{1 + \cfrac{1}{1 + \cfrac{1}{1 + \cdots}}}$$

Continued Fraction
(discoverer unknown)

$$\frac{4}{\pi} = 1 + \cfrac{1^2}{2 + \cfrac{3^2}{2 + \cfrac{5^2}{2 + \cdots}}}$$

Continued Fraction
(William Brouncker, 1655)

$$\frac{2}{\pi} = \frac{\sqrt{2}}{2} \cdot \frac{\sqrt{2 + \sqrt{2}}}{2} \cdot \frac{\sqrt{2 + \sqrt{2 + \sqrt{2}}}}{2} \cdots$$

Infinite Product
(François Viète, 1593)

$$\frac{\pi}{2} = \frac{2 \cdot 2 \cdot 4 \cdot 4 \cdot 6 \cdot 6 \cdots}{1 \cdot 3 \cdot 3 \cdot 5 \cdot 5 \cdot 7 \cdots}$$

Infinite Product
(John Wallis, 1655)

$$\frac{\pi}{4} = \frac{1}{1} - \frac{1}{3} + \frac{1}{5} - \frac{1}{7} + - \cdots$$

Infinite Series
(James Gregory, 1671)

$$\frac{\pi^2}{6} = \frac{1}{1^2} + \frac{1}{2^2} + \frac{1}{3^2} + \frac{1}{4^2} + \cdots$$

Infinite Series
(Leonhard Euler, 1736)

$$e = 1 + \frac{1}{1} + \frac{1}{1 \cdot 2} + \frac{1}{1 \cdot 2 \cdot 3} + \frac{1}{1 \cdot 2 \cdot 3 \cdot 4} + \cdots$$

Infinite Series
(Leonhard Euler, 1748)

given in 1873 by Charles Hermite (1822–1901). The transcendence of π turned out to be much more difficult to prove, and was finally established in 1882 by Ferdinand Lindemann (1852–1939). It settled once and for all the age-old problem of "squaring the circle," i.e., constructing, by straightedge and compass alone, a square whose area equals the area of a given circle. The transcendence of π meant that the construction could not be done, thus "solving" the problem in the negative. Later, in 1874, Georg Cantor (1845–1919) showed that even though there are infinitely many algebraic numbers and infinitely many transcendental numbers, the latter are actually more numerous than the former, showing for the first time that there exist different types of infinities.

One of the most beautiful results in all of mathematics was discovered in 1748 by the Swiss mathematician Leonhard Euler (1707–1783). He found that the numbers 0, 1, π, e, and i (i being the "imaginary unit," the square root of -1) are all connected by the formula

$$e^{\pi i} + 1 = 0$$

This is the celebrated Euler formula; it has been regarded by many as having no less than divine powers, since it relates the fundamental constants of arithmetic (0 and 1), of geometry (π), of analysis (e), and of complex numbers (i) in one simple equation.

The three numbers π, ϕ, and e show up in many infinite series and products, some of which are given in Table 2 (see p. 51).

Cantor's New Look at the Infinite

I see it, but I don't believe it!

— Georg Cantor in a letter to
Richard Dedekind, 1877

Georg Cantor was born in St. Petersburg (now Leningrad) on March 3, 1845. His parents had emigrated from Denmark; his father was a converted Protestant and his mother a born Catholic, but according to some evidence both were of Jewish origin. It may be that this multicultural background played a role in Cantor's early interest in medieval theological arguments, particularly those concerning continuity and the infinite. From St. Petersburg the family moved to Germany, and it was there, at the University of Halle, that in 1874 he published his first significant work on the concept of infinity. This was only the first of a series of works to be published between 1874 and 1884, and it at once changed the entire foundation on which the concept had been based thus far.

Up until Cantor's time, infinity has always been regarded in a numerical sense, as a kind of number larger than all numbers. Of course, since every number can be succeeded by an even larger number, there simply is no such thing as *the* largest number. Still, the essense of the infinite was its association with the very large— or the very small.

Furthermore, ever since Aristotle's time mathematicians have made a careful distinction between what they called the *potential* infinite and the *actual* (or "completed") infinite. The former involves a process that can be repeated again and again without end, but which, at any given stage, still encompasses only a finite number of repetitions. The set of natural numbers 1, 2, 3, . . . is potentially infinite because every natural number has a successor, and yet at each stage of the counting process, no matter how advanced, we have enumerated only a finite number of elements. The actual infinite, on the other hand, involves a process which has already acquired, at every stage, an infinite number of repetitions. The set of integers, when arranged in their "natural" order . . ., −3, −2, −1, 0, 1, 2, 3, . . ., comprises an actually infinite set, since at every stage there are already infinitely many integers present. Mathematicians were willing to accept the former type of infinity, yet they categorically rejected the latter. To quote Aristotle himself in his *Physics:* "The infinite has a potential existence . . . There will not be an actual infinite." And some two thousand years later, Carl Friedrich Gauss (1777–1855) expressed the same view in a letter to his friend Schumacher, dated 1831:

> I must protest most vehemently against your use of the infinite as something consummated, as this is never permitted in mathematics. The infinite is but a façon de parler, meaning a limit to which certain ratios may approach as closely as desired when others are permitted to increase indefinitely.

The limit concept itself, in other words, was regarded as a potentially infinite process. Gauss's remark was meant as a rebuke to occasional transgressors who used the notion of infinity—and even the symbol for infinity—as though it was an ordinary number, subject to the same rules of arithmetic as ordinary numbers are.[1]

Cantor dispelled these well-established views. First, he accepted the actual infinite as a full-fledged mathematical being by insisting that a set, and an infinite set in particular, must be regarded as a *totality,* as an object which our mind should perceive as a whole. But this amounts to removing the distinction between potential and actual infinities. Indeed, by arranging the integers not in their natural order but according to the sequence 0, 1, −1, 2, −2, 3, −3, . . ., we see at once that the distinction becomes meaningless.[2] Moreover, Cantor argued that to deny the actual infinite means to deny the existence of irrational numbers, for such numbers have an infinite decimal expansion, whereas any finite decimal would only be a rational approximation.

Next, Cantor showed that, again contrary to the prevailing views, there is not just one infinity, but many *classes* of infinity, classes which are essentially different from each other but which can be compared to one another in much the same way as ordinary numbers can. There exists, in other words, an entire *hierarchy*

---

[1] Even great mathematicians were not immune from this sin. Euler, for example, did not hesitate to say that $1/0$ is infinite and that $2/0$ is twice as large as $1/0$.

[2] According to the definition of set, the order in which we place the individual elements in a set is immaterial. Thus the set consisting of the letters a, b, c is the same as that consisting of b, a, c.

of infinities, and in this hierarchy one can speak of infinities that are greater than other infinities!

Needless to say, to express such bizarre views in the nineteenth century was no less than an act of rebellion, for they stood in direct opposition to the doctrines of the greatest mathematicians of the time. Cantor himself acknowledged this when in 1883 he said: "I place myself in a certain opposition to widespread views on the mathematical infinite and to oft-defended opinions on the essence of number." This was a humble reference to the fierce criticism with which his ideas were met by the mathematical community. The storm that was about to lash out exceeded in its fury anything that had been seen before, and it would ultimately lead to the final tragic years of his life.

Cantor based his conclusions on two remarkably simple concepts: that of a set and that of a one-to-one correspondence. A *set* is simply a collection of objects, such as a child's toys, the letters of the English alphabet, or the counting numbers 1, 2, 3, ... The latter is an example of an *infinite* set, one having infinitely many elements. Now take two different sets, such as the set consisting of the letters *a, b, c* and the set consisting of the numbers 1, 2, 3. We can establish a *one-to-one correspondence* (1:1 correspondence, for short) between the elements of the two sets by matching, say, the letter *a* with the number 1, *b* with 2, and *c* with 3; or we could match *a* with 2, *b* with 3, and *c* with 1. The actual way in which we pair the elements of the two sets is immaterial—as long as we can pair them one for one, so that no element in either set will be left alone. Now you may say that this is possible only because the two sets have the same number of elements. This indeed is true—for finite sets. If two *finite* sets have the same number of elements, a 1:1 correspondence between them can always be established. And the converse is just as true: if we can establish a 1:1 correspondence between two finite sets, we can conclude with absolute certainty that they have the same number of elements.

But what about infinite sets? Would it be possible, for example, to match on a 1:1 basis the set of all counting numbers with the set of all even numbers? At first thought this seems impossible, since there seem to be twice as many counting numbers as there are even numbers. And yet, if we arrange all the even numbers in a row according to their magnitude, then this very act already shows that such a matching is possible:

$$
\begin{array}{cccccccc}
2 & 4 & 6 & 8 & 10 & 12 & \cdots \\
\updownarrow & \updownarrow & \updownarrow & \updownarrow & \updownarrow & \updownarrow \\
1 & 2 & 3 & 4 & 5 & 6 & \cdots
\end{array}
$$

So our intuition was wrong! The apparent paradox in this situation is that an infinite set may be matched, element for element, with a subset of itself. Now this was not at all new in Cantor's time; already Galileo, in his *Dialogues on Two New Sciences* (1636), recognized the possibility of matching the squares of the counting numbers (that is, the numbers 1, 4, 9, 16, 25, . . .) with the counting numbers, even though there seem to be many more counting numbers than squares.[3] However, whereas Galileo merely acknowledged the absurdity of this situation without trying to resolve it, Cantor turned it around. He declared, simply, that *whenever two sets—finite or infinite—can be matched by a 1 : 1 correspondence, they have the same number of elements.* Thus, he concluded, there are just as many even numbers as there are counting numbers, just as many squares as counting numbers, and just as many integers (positive and negative) as counting numbers. This last fact can easily be seen if we arrange the integers according to the scheme 0, 1, −1, 2, −2, 3, −3, ... Again, the very fact that we can do this means that the integers can be matched with the counting numbers one for one:

$$0 \quad 1 \quad -1 \quad 2 \quad -2 \quad 3 \quad -3 \quad \cdots$$
$$\updownarrow \quad \updownarrow \quad \updownarrow \quad \updownarrow \quad \updownarrow \quad \updownarrow \quad \updownarrow$$
$$1 \quad 2 \quad 3 \quad 4 \quad 5 \quad 6 \quad 7 \quad \cdots$$

Cantor called any set that can be matched 1 : 1 with the set of counting numbers a *countable,* or *denumerable,* set. Thus, the even numbers, the odd numbers, the integers, the squares, and also the primes are all denumerable.

It thus appears as though infinite sets violate one of our most deeply rooted experiences, namely that "The whole is greater than the part."[4] But then, our experiences are by necessity confined to the finite world; they never transcend to the infinite. Why, asked Cantor, should infinite sets obey the same laws as finite sets do? After all, we have already encountered situations—infinite series, for example—where the ordinary rules of arithmetic for finite calculations break down, so we are no longer justified in

---

[3] This and other paradoxes involving the infinite were the subject of a work by Bernard Bolzano (1781–1848), a Czech theologian, philosopher, and mathematician. In a small book, *Paradoxes of the Infinite,* published posthumously in 1851 (the manuscript was completed just eighteen days before the author's death), Bolzano came very close to Cantor's ideas about infinite sets. Unfortunately, the work received very little recognition until many years after the author's death.

[4] This statement is the last of the ten postulates in Euclid's *Elements.* We will have more to say about these postulates in Chapter 16.

assuming *a priori* that infinite sets behave in the same way as finite sets do. As we have seen, the fact that an infinite set may be no larger than its own subset had already been noticed long before Cantor and had been regarded as an inexplicable paradox. It was Cantor's insight to realize that, far from being a paradox, this fact represents the most fundamental property of any infinite set: *that it can be matched, one-to-one, with a proper subset of itself.* [5] Cantor, in fact, used this very fact as the definition of an infinite set—the first time ever that this concept was defined in a clear and precise way, devoid of any mysticism or vagueness.

But let us return for a moment to the sets we have just considered—the counting numbers, the even numbers, the integers, the squares, and the primes. You may ask, is not the denumerability of these sets simply a result of the big gaps that exist between their members? There are gaps of one unit each between the counting numbers (as also between the integers), gaps of two units each between the even numbers, gaps of increasing magnitude between the squares (these gaps follow the sequence of odd numbers), and gaps of an irregular nature between the primes. So, we might conclude, to match these sets with the counting numbers is merely a process of rearranging their elements according to some rule. *Any* infinite set that has gaps of this sort can be matched on a 1:1 basis with the counting numbers, and is therefore denumerable. And conversely, it would seem that in order for a set to be denumerable, it must have such gaps between its members.

Far from it! The first of these conclusions is indeed correct, but not the second. Cantor was able to show that even sets which are "dense," where there are no such gaps between the members, may be denumerable. We have already met such a set—the rational numbers. In 1874 Cantor made the historic discovery that notwithstanding their density, the rational numbers are denumerable. Intuition, it seems, is a very poor guide when dealing with the infinite!

To show this, Cantor pointed out that we must first abandon our natural tendency to arrange any set of numbers according to their magnitude. True, we have already abandoned this tendency with the integers when we showed that they are denumerable. But now our task is more difficult precisely because of the

---

[5] A *subset* is any collection of objects taken from the original set and regarded as a new set. For example, the set {*a, b*} is a subset of the set {*a, b, c*}; so are the sets {*a, c*}, {*b, c*}, {*a*}, {*b*}, {*c*}, {*a, b, c*} and { }. The last subset is the *empty set*. A *proper* subset is any subset other than the original set itself; all the above-mentioned sets except {*a, b, c*} are proper subsets of the original set.

density of the rational numbers—the fact that between any two rationals, no matter how close, a third one can always be found. The density of the rationals would lead one to think that the task of enumerating them according to *any* scheme is doomed to fail; yet Cantor showed that there is a way to do just that— to list all the rational numbers one by one, without leaving even a single one out. Cantor's method was to arrange the rational numbers in an *infinite array:*

$$
\begin{array}{cccccc}
1/1 & 2/1 & 3/1 & 4/1 & 5/1 & \cdots \\
1/2 & 2/2 & 3/2 & 4/2 & 5/2 & \cdots \\
1/3 & 2/3 & 3/3 & 4/3 & 5/3 & \cdots \\
1/4 & 2/4 & 3/4 & 4/4 & 5/4 & \cdots \\
1/5 & 2/5 & 3/5 & 4/5 & 5/5 & \cdots \\
\end{array}
$$

The first row in this array consists of all fractions whose denominator is 1, that is, all the natural numbers; the second row consists of all fractions whose denominator is 2, and so on. (Within each row, the fractions are indeed arranged by their magnitude.) Cantor then traversed a path along this array:

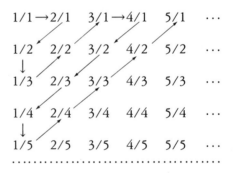

If we follow this path all along—one step to the right, then diagonally down, then one step down, then diagonally up, then again one step to the right, and so on *ad infinitum*—we will cover *all* positive fractions, one by one. True, along our path we will encounter fractions that have already been met before under a different "name," such as 2/2, 3/3, 4/4, and so on, all of which are equal to 1; these fractions we simply cross out and then continue our path as before. In this way we will have, essentially, arranged all positive fractions in a row, one by one. In other words, we can count them—they are denumerable! The discovery that the rational numbers are denumerable—in defiance of our intuition— left such a deep impression on Cantor that he exclaimed, "I see it, but I don't believe it!"

Thus, the set of fractions, despite its dense population, has just as many members as the set of counting numbers, with its big gaps. At this point, Cantor decided to give all denumerable sets a designation: he called them sets of *power* $\aleph_0$ ($\aleph$, pronounced "aleph," is the first letter of the Hebrew alphabet).[6] All sets of power $\aleph_0$ have exactly as many members as the set of counting numbers, and are therefore denumerable.

At this point we might begin to suspect (as Cantor himself originally had) that perhaps all infinite sets are denumerable, but Cantor showed that there are sets whose elements are so dense that they cannot be counted. One such set is the set of points along an infinite line, the number line. These points, in turn, correspond to our real number system, the collection of all decimals in their various forms (see p. 47). These two sets form a *continuum.* They are not denumerable; they contain more elements—vastly more—than a denumerable set. Their infinity is of a higher power than $\aleph_0$. Cantor called this type of infinity C, the infinity of the continuum.

---

[6] Cantor had originally used the German word *Mächtigkeit,* which translates into "power." Later this was replaced by the term "cardinal number."

10 Beyond Infinity

To show that the real numbers cannot be counted, Cantor first established a fact which, if anything, seems to be almost beyond belief: *There are as many points along an infinite straight line as there are on a finite segment of it.* The proof, shown in Fig. 10.1, is so simple that one wonders why no one before him had made the discovery. It shows that our conception of a line as being made up of many dots of ink is fundamentally wrong: the physical dot and the mathematical point have absolutely nothing in common!

Having shown this, all that remained for Cantor to do was to show that the points along a *finite* line segment could not be counted. He did this in his celebrated "diagonal" argument. As his finite segment, he chose the interval on the number line from 0 to 1, that is, the unit segment. Each point along this segment corresponds to a proper fraction, a decimal between 0 and 1. To avoid possible ambiguities, let us agree to regard all these decimals as non-terminating, so that a fraction such as 0.5 will be replaced by the equivalent fraction 0.4999 . . . (see p. 32).

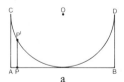

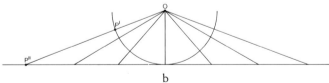

a

b

Figure 10.1. (*a*) *The points of the segment* AB *are matched one-to-one with the points of the semicircle* CD, *proving that the segment and the semicircle have the same number of points.* (*b*) *The points of the semicircle are now matched one-to-one with the points of the entire line. Thus a finite line segment and an infinite line have exactly the same number of points!*

Cantor now assumed that this set of decimals *can* be counted; this would mean that we could list all the decimals between 0 and 1, one by one, without leaving any of them out. Let us represent this list symbolically by using letters instead of actual digits, so that the first decimal will be $0.a_1a_2a_3 \ldots$, the second decimal $0.b_1b_2b_3 \ldots$, and so on until we cover *all* the decimals between 0 and 1.[1] Thus our list is:

$$r_1 = 0.a_1a_2a_3 \cdots$$
$$r_2 = 0.b_1b_2b_3 \cdots$$
$$r_3 = 0.c_1c_2c_3 \cdots$$
$$\cdots\cdots\cdots\cdots\cdots$$

Now, according to our assumption, this list includes *all* the real numbers between 0 and 1. But Cantor now "constructed" a real number between 0 and 1 which is not included in the list. He did this by selecting as the first digit of the new number any digit different from a_1, as the second digit any digit different from b_2, as the third digit any digit different from c_3, and so on *ad infinitum*. (To illustrate, if the first three numbers in our list were, say, $0.274 \ldots$, $0.851 \ldots$ and $0.308 \ldots$, we would select as the first digit any digit except 2, as the second digit any digit except 5, as the third, any digit except 8, etc.) Thus our new number has the form:

$$r = 0.x_1x_2x_3 \cdots$$

where $x_1 \neq a_1$, $x_2 \neq b_2$, $x_3 \neq c_3$, and so on. (It is from this pattern that the name "diagonal" came.) This new number is evidently a real number between 0 and 1, and yet by its very method of construction it differs from any of the numbers in the original list. And therein lies the contradiction, since we have assumed that our list includes *all* the real numbers between 0 and 1. Thus the assumption that the real numbers in this interval (and therefore along the entire number line) are denumerable has been proven to be wrong. The real numbers, therefore, form a non-denumerable set, a number continuum, whose power Cantor denoted by the letter C.[2]

---

[1] The fact that we are using the letters of the English alphabet does not, of course, mean that our list should come to an end after the 26th number. In fact, we could have used a more general notation such as a double subscript for the digits: $0.a_{11}a_{12}a_{13} \ldots$ for the first decimal, $0.a_{21}a_{22}a_{23} \ldots$ for the second, and so on. For the sake of simplicity, however, we will use the single subscript notation.

[2] Since the real numbers are composed of the rational numbers and the irrational numbers, of which the former are denumerable, the non-denu-

Thus Cantor created, in effect, a *hierarchy* of infinities, in which all sets of power C stand higher than the sets of power $\aleph_0$, the denumerable sets. There is no infinite set of power less than $\aleph_0$; even if we deplete the set of counting numbers by removing from it, say, the even numbers, we still get a set of power $\aleph_0$ (namely, the odd numbers). The same is true if we remove from the counting numbers all the odd numbers (leaving the even numbers), all composite numbers (leaving the primes), and so on; the remaining sets will still have as many elements as the original set!

On the other hand, Cantor was able to find sets whose powers are greater even than C. He did this by showing that the *set of all subsets* of a given set always has more elements than the original set. For a finite set this is quite obvious: the set $\{a, b, c\}$, for example, has the subsets $\{a\}$, $\{b\}$, $\{c\}$, $\{a, b\}$, $\{a, c\}$, and $\{b, c\}$, to which we must add the so-called "empty set" $\{\ \}$ and the set $\{a, b, c\}$ itself. Thus from the original set of three elements we derive a new set of eight ($=2^3$) elements. This result can be generalized to any finite set: if the set has n elements, then the set of all its subsets (including the empty set and the original set itself) has 2^n elements, which is always greater than n. Cantor now extended these results to infinite sets: he introduced the notion—quite revolutionary for its boldness—that from any *infinite* set one may create a new set which actually has more elements than the original set, by considering all possible subsets of the original set. This is a truly mind-boggling idea, for how can one think of something more numerous than the continuum, the set of all points of the line? We must realize—and Cantor pointed this out—that we deal here largely with a thought process, with the ability of our mind to *conceive* such sets; whether such sets actually exist (in the physical sense) is quite irrelevant to the issue.[3]

And now Cantor crowned it all with the inevitable conclusion: If from any given set—finite or infinite—one can create a new set which has more elements than the original one ("more" in

---

merability of the reals means that the irrational numbers, too, are non-denumerable. But this implies that the "holes" left along the number line by the fact that some points correspond to irrational numbers are actually more numerous than the "non-holes" of the rationals! This is yet another paradox involving infinity which our finite intuition finds hard to accept.

[3] One might think that the two-dimensional plane should have more points than the one-dimensional line, but here again our intuition fails: Cantor showed that the plane has just as many points as the line. The same is true of three-dimensional space, and in fact of any "hyper-space" having a denumerable number of dimensions. Dimensionality thus has nothing to do with the number of points a space contains.

the ordinary sense of the word for finite sets, and in the sense of power for infinite ones), then this process can be repeated; that is, from the new set we can create a set with still more elements, and so on *ad infinitum!*[4] We have generated, in effect, an *infinite hierarchy of infinite sets,* in which each new set (of subsets) has a greater power than the one from which it was derived. Cantor designated these increasing powers by $2^{\aleph_0}$, $2^{2^{\aleph_0}}$, and so on—in analogy with the finite case, where the powers follow the sequence 2^n, 2^{2^n}, etc. In this hierarchy, all sets of the same power can be matched with one another on a $1:1$ basis, while sets of different powers cannot be so matched.

The powers $\aleph_0$, $2^{\aleph_0}$, $2^{2^{\aleph_0}}$, . . . came to be known as *transfinite cardinal numbers.*[5] They extend to the infinite case the numerical hierarchy for finite sets, $n < 2^n < 2^{2^n} < . . .$, so that we are able to write $\aleph_0 < 2^{\aleph_0} < 2^{2^{\aleph_0}} < . . . .$ Cantor now extended this analogy one step further by creating an *arithmetic of transfinite cardinals.* This arithemtic looks very strange to our "common sense," used as we are to the ordinary arithmetic of finite numbers. In transfinite arithmetic we find such bizarre rules as $\aleph_0 + \aleph_0 = \aleph_0$ (which merely says that if we combine two denumerable sets, the united set will still be denumerable), $\aleph_0 \cdot \aleph_0 = \aleph_0$ (which says that the union of a denumerably infinite number of denumerable sets is still denumerable),[6] and other similarly strange formulas.

With these developments, the long and arduous struggle to clarify and demystify the concept of infinity seemed finally to have come to its successful conclusion. "The solution of the difficulties which formerly surrounded the mathematical infinite," said Bertrand Russell in 1910, "is probably the greatest achievement of which our age has to boast." And David Hilbert (1862–1943), one of the greatest mathematicians of our time, had this to say in Cantor's tribute: "No one shall drive us from the paradise Cantor created for us." Yet Cantor's personal life soon took a tragic turn. Towards the end of his life he suffered from spells

---

[4] Even for a finite set it is quite remarkable how fast the number of elements grows in this process. If we begin with a set of three elements, the set of its subsets will have $2^3 = 8$ elements, the set of subsets of this set will have $2^8 = 256$ elements, the next set will have 2^{256} elements (which is aproximately 10^{77}, a huge number), and so on. The immensity of this last number can perhaps be grasped from the fact that the total number of stars in the universe is estimated at only 10^{22}.

[5] In set theory, the ordinary counting numbers are referred to as the finite cardinal numbers.

[6] We have tacitly assumed this rule in showing that the rational numbers are denumerable (p. 58).

of depression which increasingly interfered with his creative work. This was at least partially caused by the severe criticism with which his ideas were met by many of his colleagues. He was particularly attacked by his former teacher, Leopold Kronecker (1823–1891), himself a mathematician of fame but an ultraconservative who rejected from mathematics not only the actual infinite but everything which was not directly based on the natural numbers.[7] Yet one suspects that Kronecker's vehement attacks on Cantor were not purely academic; they had undertones of jealousy toward his onetime student who suddenly outshone him in fame. Cantor died in a mental institution in 1918.

Nor would the paradise Hilbert spoke of last for long. In a historic address before the Second International Congress of Mathematicians, held in Paris in the opening year of the new century, Hilbert challenged the mathematical community with a list of 23 unsolved problems whose solution he regarded as of the utmost importance to the future of mathematics. The very first among his problems referred to a question which Cantor himself had already raised in 1884. As we have seen, Cantor created a hierarchy of infinities represented by the transfinite cardinals $\aleph_0$, $2^{\aleph_0}$, $2^{2^{\aleph_0}}$, ... But he also showed that the real numbers have a transfinite cardinal C which is greater than $\aleph_0$, and in fact he was able to prove that $2^{\aleph_0}$ is *equal* to C, i.e., that the set of all subsets of the natural numbers has exactly as many elements as the set of all real numbers. The question which then presented itself to Cantor was, can one find a set with a power *between* $\aleph_0$ and C? Cantor conjectured that the answer is "no," but he was unable to prove his conjecture.

An analogy with finite sets would be in place here. Beginning with a set of two elements ($n = 2$), we obtain, by repeatedly constructing sets of subsets, the sequence of powers $2^2 = 4$, $2^4 = 16$, $2^{16} = 65,536$, and so on. But we can never get in this process a set of, say, three elements (this is true even if we start with any other value of n). Thus the sets of subsets represent only a small fraction of all possible finite sets. Cantor had at first assumed that a similar situation holds for infinite sets, i.e., that there exists sets with a power, say, between $\aleph_0$ and $2^{\aleph_0}$ (just as

---

[7] It was Kronecker who coined the famous saying, "God made the integers; all the rest is the work of man." His uncompromising rejection of the actual infinite led him so far as to ban irrational numbers from mathematics, in effect bringing it back to the state it had in Pythagoras's time.

$2 < 3 < 2^2$). But when all his attempts to "find" such a set failed, he began to suspect that no such sets exist; more specifically, that no sets exists whose power is greater than $\aleph_0$ but smaller than $2^{\aleph_0}$ (and therefore C). Cantor's conjecture came to be known as the *Continuum Hypothesis.* [8] In his historic address in the year 1900, Hilbert challenged mathematicians to either prove the Continuum Hypothesis or disprove it by a counterexample, i.e., to produce a set with a cardinality between $\aleph_0$ and C.

The challenge of proving or disproving the Continuum Hypothesis would loom over the mathematical community for the next 60 years. When the question was finally settled in 1963, the solution was something of a surprise: the hypothesis turned out to be both true and false—depending on what assumptions one starts from! This startling discovery sent shock waves through the mathematical world whose effects are still being felt today.[9] It showed that the Continuum Hypothesis is independent of the axioms of set theory; it can be regarded as an additional axiom which we are free to accept or reject. Yet surprising though it was, this idea was not without a precedent, for a similar notion had already found its way into another branch of mathematics some hundred years before. This branch was geometry, and it is to it that we now turn.

---

[8] More generally, the conjecture says that no transfinite cardinal exists between $\aleph_0$ and $2^{\aleph_0}$, between $2^{\aleph_0}$ and $2^{2^{\aleph_0}}$, and so on. This is known as the *Generalized Continuum Hypothesis.*
[9] We will briefly discuss some of these developments in the Appendix.

Geometric Infinity   Part II

*Infinity is where
things happen that
don't.*

— an anonymous
schoolboy

*The serpent biting its own
tail as a symbol of eternity.
Reprinted from Bruno
Murari, Discovery of
the Circle, Wittenborn
and Company, New York,
1970, with permission of
Wittenborn Art Books, Inc.*

*Musical notation without
beginning or end for a
sound-giving object.
Reprinted from Bruno
Murari, Discovery of
the Circle, Wittenborn
and Company, New York,
1970, with permission of
Wittenborn Art Books, Inc.*

11 Some Functions and Their Graphs

The range of focus of your telescope is from 15 feet to infinity and beyond.

— from the manual of a telescope manufacturer

Geometry is the study of form and shape. Our first encounter with it usually involves such figures as triangles, squares, and circles, or solids such as the cube, the cylinder, and the sphere. These objects all have finite dimensions of length, area, and volume—as do most of the objects around us. At first thought, then, the notion of infinity seems quite removed from ordinary geometry. That this is *not* so can already be seen from the simplest of all geometric figures—the straight line. A line stretches to infinity in both directions, and we may think of it as a means to go "far out" in a one-dimensional world. As we shall see, it was this simple idea that gave rise, around the middle of the nineteenth century, to one of the most profound revolutions in mathematical thought—the creation of non-Euclidean geometry.

Next to the straight line in simplicity is the circle. Although finite in size, one can travel around it endlessly, always covering the same ground. It is for this reason that we regard the circle (and its three-dimensional counterpart, the sphere) as being *finite and yet unbounded.* Since times immemorial the circle has been the symbol of regular recurrence, of periodicity, and of perpetual motion—one only has to think of the diurnal cycle of day and night, the annual cycle of the seasons, or the eternal cycle of life and death. Can one relate the actual infinity of the straight line to the symbolic infinity of the circle? Yes indeed; as we will see, there is a certain mathematical transformation that transforms the one into the other.

In elementary algebra we learn about functions and their graphs.

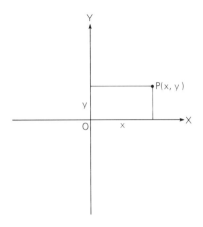

Figure 11.1. *Rectangular coordinates.*

A *function* is a relation between two or more variables.[1] We can express such a relation by a mathematical equation, but it is often more convenient to use an idea first proposed in 1637 by René Descartes (1596–1650): to plot the function graphically in a coordinate system. Usually our coordinate system consists of two perpendicular lines, or "axes," with scales marked along them. A point P in the plane is determined by its distances x and y from these axes; we say that P has the *coordinates* x and y, written as (x, y) (Fig. 11.1). Now if y is a function of x, then as we let x vary, the point P will trace a curve, the graphical representation of the function. For example, the equation $2x + 3y = 1$ (or more generally, $Ax + By = C$) is represented by a straight line, $x^2 + y^2 = 1$, by the unit circle, and $2x^2 + 3y^2 = 1$ (or more generally, $Ax^2 + By^2 = 1$, where A and B are positive constants), by an ellipse (Figs. 11.2–11.4).

The circle and the ellipse both have finite dimensions—they occupy finite regions of the coordinate plane. Other curves extend to infinity in one or more directions; we have already seen the straight line, to which we now add the parabola and the hyperbola (Figs. 11.5 and 11.6). Taken together, these five curves form the family of *conic sections,* so called because they can all be obtained by cutting a cone with a plane at various angles of incidence (Fig. 11.7). All five conic sections were known to the Greeks, and were later destined to play a key role in astronomy, for it

---

[1] Strictly speaking the definition of function requires that each permissible value of one variable should produce exactly one value of the other variable.

11. Some Functions and Their Graphs

Figure 11.2. *A straight line:* $Ax + By = C$.

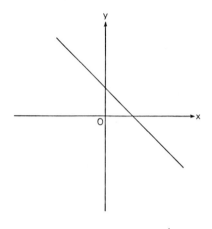

Figure 11.3. *The unit circle:* $x^2 + y^2 = 1$.

Figure 11.4. *The ellipse* $Ax^2 + By^2 = 1$ $(A, B > 0)$.

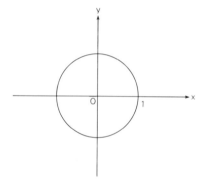

Figure 11.3

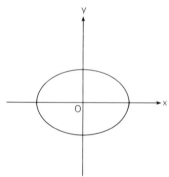

Figure 11.4

Figure 11.5. *The parabola* $y = ax^2$ $(a > 0)$.

Figure 11.6. *The hyperbola* $y = a/x$ $(a > 0)$.

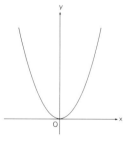

Figure 11.5

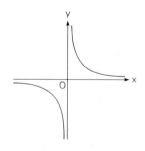

Figure 11.6

70

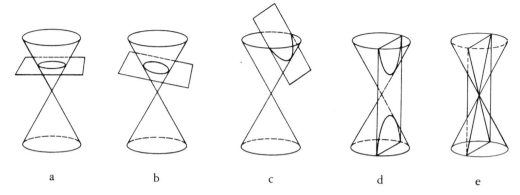

a   b   c   d   e

Figure 11.7. *The five conic sections.*

is along these curves that any celestial body—be it a planet, a comet, or a satellite—must move.

The parabola, in particular, is of interest to us here. It is the locus of all points equidistant from a given line *d* and a point *F* not on *d* (Fig. 11.8). The line *d* is the *directrix,* and the point *F* the *focus,* of the parabola. Like all conic sections, the parabola is a symmetric curve; its line of symmetry, called the *axis,* is the line perpendicular to the directrix and passing through the focus. If we think of the parabola as a reflecting surface, and if we place a source of light exactly at the focus, then the rays of light emanating from this source will all be reflected in one direction, namely, parallel to the axis. And conversely, rays of light arriving at the parabola from infinity in a direction parallel to the axis will converge at the focus (Fig. 11.9). This unique property of the parabola is used in our modern cars, whose headlights have a parabolic cross-section and therefore reflect light with great efficiency. But apparently the idea is not new: legend has it that Archimedes, who was well acquainted with the conic sections, used it to defend the city of Syracuse against the Roman invaders in their war against

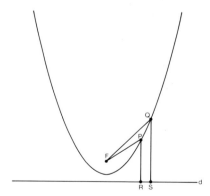

Figure 11.8. *The parabola is the locus of all points equidistant from the focus* F *and the directrix* d.

71

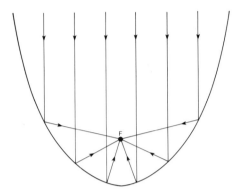

Figure 11.9. *The reflecting property of the parabola: all rays of light arriving from infinity along the axis converge at the focus* F.

Carthage. He is said to have built huge parabolic mirrors which he aimed at the Roman fleet besieging the city, so that the sun's rays, converging at the focus of each parabola, set the enemy's ships ablaze.[2]

The parabola, like the straight line, is a *continuous* curve: it can be drawn with one stroke of a pencil, so to speak. This is not so with the hyperbola, which consists of two separate and diverging branches. You can go to infinity along one branch and to negative infinity along the other, but you can never switch branches: it is as though the hyperbola disappears from sight at infinity, only to reemerge again from negative infinity. This eccentric behavior—quite common in mathematics—must have left many a layman with a profound sense of bewilderment, as noted humorously by Winston Churchill in *My Early Life:*

> *I saw, as one might see the transit of Venus, a quantity passing through infinity and changing its sign from plus to minus. I saw exactly how it happened and why the tergiversation was inevitable . . . but it was after dinner and I let it go!*

The hyperbola is associated with a pair of straight lines that "show the way to infinity." These lines, called the *asymptotes,* are the tangent lines to the hyperbola at infinity. The curve approaches these lines ever closer, but never touches them. The asymptotes are thus the graphical equivalent of the limit concept discussed in Chapter 3.

Another function with an asymptote is the *exponential function* $y = a^x$, where the constant a is any number greater than 1. This function represents a continuous increasing geometric progression (see p. 29): if the "independent variable" x is increased in equal

---

[2] The word "focus" in Latin means "fireplace."

amounts, then the "dependent variable" y increases in equal ratios. The graph of this function grows moderately at first, then at an increasing rate to infinity. In the same manner, the function $y = a^{-x}$ represents a continuous *decreasing* geometric progression: its graph "decays" asymptotically to zero. The two graphs, shown side by side in Fig. 11.10, are exact mirror images of each other.

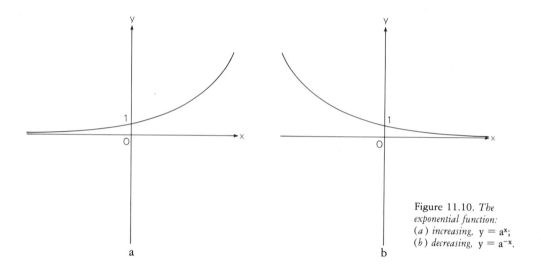

a

b

Figure 11.10. *The exponential function:* (*a*) *increasing,* $y = a^x$; (*b*) *decreasing,* $y = a^{-x}$.

The rectangular coordinate system of Descartes is not the only way to plot a function. We may use instead *polar coordinates,* in which the independent variable is a rotation through an angle θ around a fixed point, called the *pole,* while the dependent variable is the distance r from this pole (Fig. 11.11). Plotting a function in polar coordinates may alter its familiar graph beyond recogni-

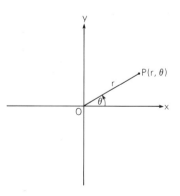

Figure 11.11. *Polar coordinates.*

Figure 11.12. *A linear spiral, also known as the Archmidean spiral:* r = aθ.

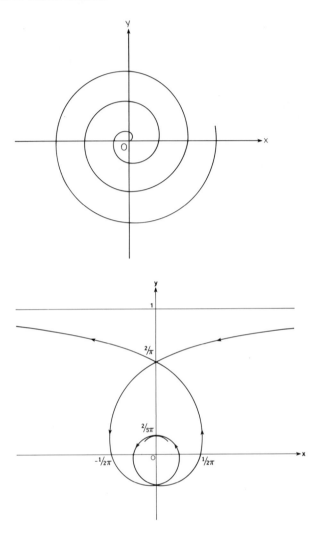

Figure 11.13. *A hyperbolic spiral:* r = 1/θ.

tion: a straight line becomes a linear spiral (Fig. 11.12),[3] the hyperbola becomes a hyperbolic spiral (Fig. 11.13), and the exponential function a *logarithmic spiral* (Fig. 11.14). This last curve, with its graceful shape, has won the admiration of many scholars, not only because of its frequent occurrence in nature but also due to its unusual mathematical properties. Of these I will mention three.

---

[3] Also known as the Archimedean spiral; it has the shape of a rope tightly wound around a pole.

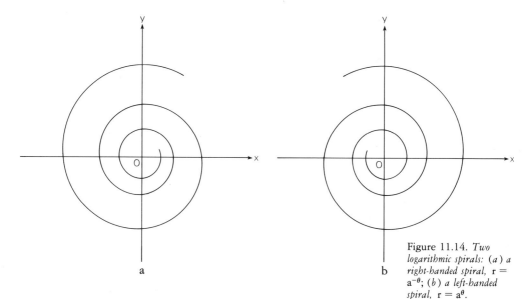

a   b

Figure 11.14. *Two logarithmic spirals: (a) a right-handed spiral,* $r = a^{-\theta}$; *(b) a left-handed spiral,* $r = a^{\theta}$.

First, each full rotation of the spiral increases its distance from the pole by a fixed ratio; this ratio differs from one spiral to another and determines its rate of growth.[4] Secondly, every straight line through the pole intersects the spiral at the same fixed angle (Fig. 11.15).[5] This makes the logarithmic spiral a close relative of the circle, for which the angle of intersection is 90°. (Indeed, the circle is a special logarithmic spiral whose rate of growth is zero.) Thirdly, if one traces the spiral inward towards its pole, one has to follow infinitely many turns—and yet the total distance covered is finite! This remarkable fact, shown in Fig. 11.16, was first discovered in 1645 by Evangelista Torricelli (1608–1647), a disciple of Galileo who is better known for his experiments in physics. It was, however, Jacques Bernoulli (1654–1705), the first of the Bernoulli brothers to achieve mathematical prominence, who explored this curve systematically and discovered most of its properties. He dubbed it "spira mirabillis" and requested that a logarithmic spiral be engraved on his tombstone with the inscription,

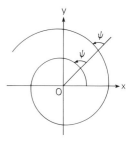

Figure 11.15. *The equiangular property of the logarithmic spiral: all rays emanating from the center O meet the spiral at the same angle* ψ.

---

[4] The logarithmic spiral thus unfolds in a geometric progression. This follows from the properties of the exponential function $r = a^{\theta}$ discussed above.

[5] It is for this reason that the logarithmic spiral is also known as the *equiangular spiral*. Thus any portion of the spiral is similar in shape to any other portion having the same angular width. It is probably this feature that is responsible for the frequent occurrence of the spiral in nature—the nautilus shell, the sunflower, and the spiral galaxies are but three examples.

Figure 11.16.
Rectification of the logarithmic spiral: the arclength of the spiral from T *to the center* O *equals the length of the tangent line from* T *to* S.

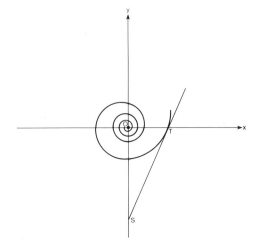

Eadem mutata resurgo ("though changed, I shall arise the same") (Fig. 11.17). These words sum up the most remarkable feature of the spiral: its invariance under most of the common transformations of geometry. Stretch it, shrink it, or rotate it—you will always end up with the same spiral!

Towards the end of the nineteenth century—just when it seemed that the concepts of continuity and infinity were finally being fully understood—several discoveries were made that cast new doubts about these concepts. We have already seen the revolution brought about by Cantor's ideas about infinite sets. Less revolutionary—though no less dramatic in its consequences—was the discovery of a new class of curves which seemed to defy our intuitive notion of geometric continuity. Dubbed "pathological" by the mathematicians, these curves gave rise to some of the most remarkable paradoxes of geometric infinity. The first of these curves was suggested in 1890 by the Italian Giuseppe Peano (1858–1932): Take a square and divide it into four equal smaller squares, then join the centers of the small squares by straight line segments, beginning at the upper-left square and ending at the lower-left one (Fig. 11.18a). Next divide each small square into four equal squares (there are now $4^2 = 16$ squares), and join their centers as in Fig. 11.18b. This process can be continued indefinitely, at least in principle; Fig. 11.18d shows the result after six steps, when the original square has been divided into $4^6 = 4,096$ small squares. As we continue in this way, the curve gradually occupies more and more of the original square, until one gets the illusion that the entire area of the square is uniformly filled up by the

Figure 11.17. *Jacob (Jacques) Bernoulli's tombstone in Basel, Switzerland. His wish that a logarithmic spiral be engraved on his tomb was fulfilled, but the engraver used the wrong spiral—an Archimedean instead of a logarithmic! Reproduced with permission from Birkhäuser Verlag AG, Basel.*

curve. The limit of this process, as the number of steps tends to infinity, is known as the Peano or "space-filling" curve.[6] Its endless meanderings carry it through every point inside the square, and its total length is infinite.

Another famous pathological curve is the "snowflake" or Koch curve, named after its inventor, the Swede Helge von Koch (1870–1924), who proposed it in 1904: We begin with an equilateral triangle of unit sides. We then trisect each side, construct

---

[6] Often named after David Hilbert (1862–1943), who investigated it in 1891.

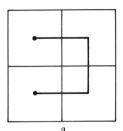
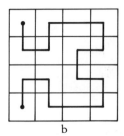

| a | b | c | d |

Figure 11.18. *The Peano curve: the unit square is divided into (a) 4 equal small squares, (b) $4^2 = 16$ small squares, (c) $4^3 = 64$ squares, and (d) $4^6 = 4,096$ squares, and the midpoints connected as shown. In the limit we get a space-filling curve that passes through every point of the original square. Its total length is infinite. Part (d) reprinted from* Mathematics in the Modern World: Readings from Scientific American, *W.H. Freeman, 1968, with permission. Originally appeared in "Geometry and Intuition" by Hans Hahn. Copyright © 1954 by Scientific American, Inc. All rights reserved.*

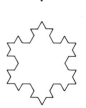

an equilateral triangle over the middle segment, and then remove each middle segment (Fig. 11.19). The result is a Star-of-David-like figure whose total length is 4/3 that of the original triangle. We now repeat this process with each of the twelve sides of the new figure, obtaining a 48-sided figure whose length is 4/3 that of the previous one, and therefore $(4/3)^2 = 16/9$ that of the original triangle. If we continue this process indefinitely, we get the strange crinkly curve shown in Fig. 11.20. Like the Peano

Figure 11.19. *The "snowflake" or Koch curve—the first three steps.*

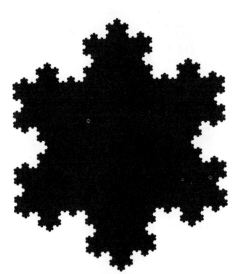

Figure 11.20. *The Koch curve in the limit: not only is its total length infinite, but the distance between any two points on the curve is infinite! Reprinted from Benoit B. Mandelbrot,* Fractals—Form, Chance, and Dimension, *W.H. Freeman, San Francisco, 1977, with the author's permission. (Figure also appears in* The Fractal Geometry of Nature *by same author, W.H. Freeman, 1982.)*

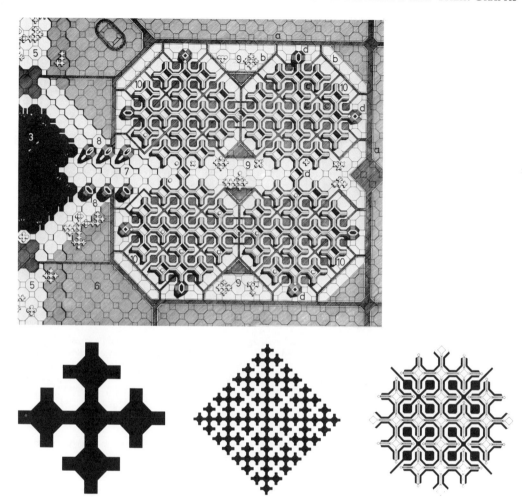

Figure 11.21. *Zvi Hecker's plan for the center of the city of Ashdod, Israel. The design resembles the construction of the Koch curve and suggests the possibility of endless continuation. Reprinted from Zvi Hecker,* Polyhedric Architecture, *The Israel Museum, Jerusalem, 1976, with the permission of the author and the Israel Museum.*

curve, it consists entirely of straight-line segmens, so that the curve is nowhere "smooth"—it nowhere has a definite direction.[7] Moreover, not only does its total length tend to infinity (as follows from the fact that at each stage the length increases by a factor of 4/3), but the distance *between* any two points on the curve, no matter how close they seem to be, becomes infinite! And yet,

---

[7] Mathematically speaking, this means that the curve nowhere has a tangent line—it can nowhere be differentiated.

79

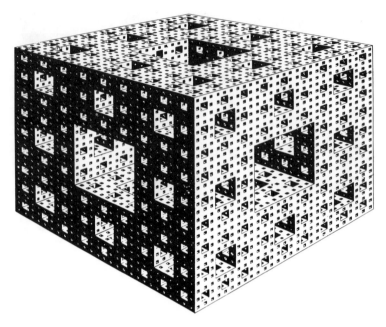

Figure 11.22. *The "Menger sponge": its volume vanishes, while the area surrounding its holes is infinite. Like the Koch curve, it can only be realized as a limiting case. Each of its external faces is known as a "Sierpinski carpet," named after the Polish mathematician Waclaw Sierpinski (1882–1969): its area vanishes, while the total perimeter of its holes is infinite. Reprinted from Leonard M. Blumenthal and Karl Menger,* Studies in Geometry, *W.H. Freeman, San Francisco, 1970, with permission.*

the figure always encompasses a finite area, so that we have in effect a finite region surrounded by infinite boundaries.[8]

The discovery of such bizarre geometric creations confronted the late nineteenth century mathematicians with a real challenge, for it shattered their intuitive notion that a continuous curve can always be traced with a steady motion of a pencil, without making any abrupt breaks. It was yet another remainder that in mathematics—and especially in dealing with infinite processes—intuition

---

[8] In his book *Fractals: Form, Chance, and Dimension,* the Polish-born American mathematician Benoit B. Mandelbrot suggests an interesting problem related to these pathological curves. He asks, "How long is the coast of Britain?", and notes that a truly rugged coastline is so irregular as to make any conventional estimate of its length (i.e., by dividing it into many straight-line segments) impossible. He concludes: "The final estimated length is not only extremely large but in fact so large that it is best considered infinite." Nevertheless, he proposes a way to measure such a length, by introducing the notion of a *fractal*—a fractional dimension.

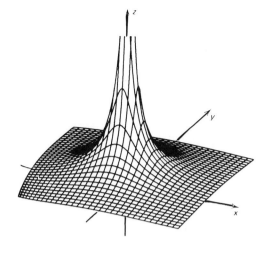

Figure 11.23. *The surface of* z = *ln* (x² + y²), *where ln denotes the natural logarithm. This surface represents a function of two independent variables. As* x *and* y *approach 0, the function becomes infinite. From* Calculus and Analytic Geometry *by Al Shenk. Copyright © 1977 by Scott, Foresman and Company. Reprinted by permission.*

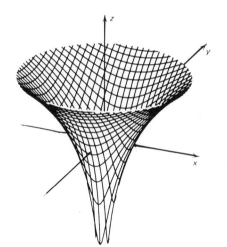

Figure 11.24. *The surface of* $z = \dfrac{1}{\sqrt{x^2 + y^2}}$: *another example of a function that becomes infinite at the origin. From* Calculus and Analytic Geometry *by Al Shenk. Copyright © 1977 by Scott, Foresman and Company. Reprinted by permission.*

is a very unreliable guide. It forced mathematicians to set new and higher standards of rigor for their professions, and indeed the very definition of the concept of curve had to be revised to accommodate these pathological cases.[9]

At the root of these paradoxes lies once again the fact that when dealing with the infinite, the part may be as large as the

---

[9] See the article "Geometry and Intuition" by Hans Hahn, *Scientific American,* April 1954.

whole. We have seen this with infinite sets, and we see it again here: take any portion of Peano's curve and magnify it as much as you please—it will look exactly like the original curve. Observe it under the most powerful microscope, and it will not resolve itself into its elements any more than the original curve did. The part is an exact replica of the whole![10]

---

[10] The Education Development Center for the Topology Films Project has produced a film called "Space Filling Curves" which shows in a dramatic way some of these features. Using computer animation, a curve is being continually magnified, while new portions of it seem to arrive at the observer from nowhere, all having the exact same shape as the entire curve.

Some Geometric Paradoxes Involving Infinity

In Part I we looked at some of the paradoxes associated with the concept of infinity. These paradoxes involved numerical quantities, e.g., sets of real numbers, infinite series, etc. Similar paradoxes also arise in geometry, where shape and form take the place of number and quantity. Some of these paradoxes are easily explained, while others touch upon some of the most fundamental notions on which geometry rests, particularly the notion of continuity.

Consider a square of unit side, divide it into four equal small squares, and shade the upper-right of these squares, as in Fig. 1a. The shaded area is one-fourth that of the original square (whose area is 1). Now divide each of the remaining, unshaded squares into four equal squares, and shade the upper-right square in each (Fig. 1b). The total shaded area is now equal to $1/4 + 3/16$, or $7/16$. Continuing in this manner, will the shaded area approach a limit? If so, what is this limit?

It is easiest to answer these questions by considering the *unshaded* area in each step. In the first step there are three unshaded squares, each with an area equal to $1/4$; thus the total unshaded area is $3/4$. In the second step there are nine unshaded squares, each with an area of $1/16$, making the total unshaded area $9/16$, or $(3/4)^2$. In the third step there will be 27 unshaded squares, each with an area of $1/64$; thus the total unshaded area will be $27/64$, or $(3/4)^3$. Continuing in this way for n steps, we get the sequence $3/4, (3/4)^2, (3/4)^3, \ldots, (3/4)^n$. This is a geometric progression with a common ratio of $3/4$; as n grows, the terms of this progression diminish and tend to 0, so that the unshaded

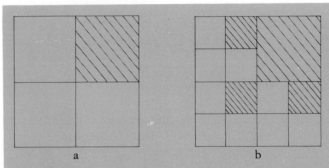

Figure 1. *The shaded square paradox: Start with a unit square, divide it into four equal small squares, and shade the upper right of these, as in (a). Then repeat the process with the remaining (unshaded) squares, as in (b). Continuing this process indefinitely, the total shaded area approaches the area of the original square. This paradox can easily be explained using a geometric progression.*

area becomes less and less significant. Therefore, the *shaded* area must approach the limit 1—the area of the original square. In other words, after sufficiently many steps, the shaded part will nearly cover the entire square, even though in each step we leave three quarters of each square unshaded!

The next paradox is even more remarkable. Consider the function $y = 1/x$, whose graph is the hyperbola shown in Fig. 2a (only the branch for positive values of x is shown). Now imagine

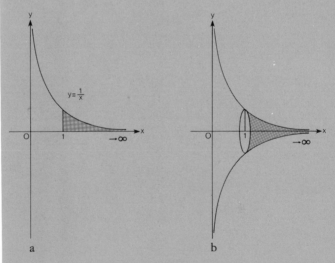

Figure 2. *The part of the hyperbola $y = 1/x$ from $x = 1$ to infinity [see (a)], when revolved about the x-axis, forms the surface of revolution shown in (b). Using calculus, one can show that the volume enclosed by this surface is finite, yet its surface area is infinite. In other words, one can fill the interior with a finite amount of paint, but an infinite amount of paint is needed to coat the surface! This paradox does not have an "elementary" explanation.*

84

that we revolve this graph about the x-axis, obtaining a solid known as a hyperboloid of revolution (Fig. 2b). Using techniques from the calculus, one can show that the surface area of this solid, taken from $x = 1$ to infinity, is infinite. More precisely, if we calculate the surface area from $x = 1$ to some value of x greater than 1, say $x = t$, and then let $t \to \infty$, the area will grow without bound. On the other hand, the *volume* of this solid from $x = 1$ to $x = t$ will approach a definite limit as $t \to \infty$; in other words, the volume is finite, even though the solid extends to infinity. Now imagine that we want to coat the outside surface of this solid with a thin, uniform layer of paint. We would never be able to complete the job, since an infinite amount of paint would be needed. Yet if we were to fill the entire inside space of the hyperboloid with paint, a finite amount would suffice! This paradox has no elementary "explanation"; it shows again that when the infinite is involved, our common sense may fail us.[1]

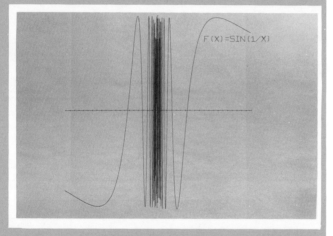

F(X)=SIN(1/X)

Figure 3. *The graph of the function* y = sin (1/x), *where sin is the sine function studied in trigonometry. The values of this function always stay between 1 and −1, but the frequency of the oscillations becomes infinite at* x = 0. *The plotter that drew this graph could not "hit" the peaks beyond a certain point—hence the irregularities near* x = 0.

Some paradoxes involve the "pathological functions" mentioned in Chapter 11. The function $y = \sin (1/x)$, for example, has the peculiar property that as x approaches 0, its graph oscillates at an ever-increasing frequency (Fig. 3), so that one can never completely plot the function. This, of course, is not particularly

[1] For a full discussion of this paradox, see Philip Gillett, *Calculus and Analytic Geometry*, D.C. Heath, Lexington, Massachusetts, 1984, p. 370.
[2] Strictly speaking, this function is undefined for $x = 0$, but we can remove this "singularity" by assigning it the value 0 there. This definition would preserve the continuity of the function at $x = 0$, since $\lim\limits_{x \to 0} x \sin (1/x) = 0$.

Figure 4. *The graph of* y *= x sin (1/x). Here the amplitude (width) of the oscillations tends to zero as* x → 0, *thus removing the "singularity" of the previous graph.*

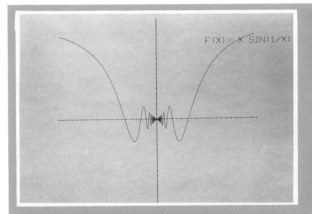

unique to our function—the hyperbola also has a "break" at $x = 0$—but unlike the hyperbola, here no part of the graph ever goes to infinity; it is the *frequency* that becomes infinite. On the other hand, if we consider the related function $y = x \sin (1/x)$, the "singularity" at $x = 0$ is removed and the function becomes continuous (Fig. 4).[2]

Finally, consider the function $y = f(x)$ defined as follows: If x is a rational number, $y = 1$; if x is irrational, $y = -1$. The graph of this function seems to consist of two horizontal lines, as in Fig. 5. But each of these lines is "punctured" by infinitely many holes—a result of the fact that between any two rational

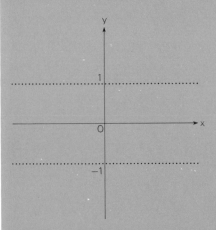

Figure 5. *The function* $y = \begin{cases} 1, \text{ if x is rational} \\ -1, \text{ if x is irrational} \end{cases}$. *Its graph consists of what appears to be two horizontal lines, but each line is broken by infinitely many holes. Moreover, since the irrational numbers are more numerous than the rationals, the upper line has more holes than the lower. This is an example of a "pathological function"—one that defies our common sense of continuity and smoothness.*

numbers, no matter how close, an irrational number can be found, and vice versa. To make the situation even more bizarre, the lower line has only a countable number of holes, while the upper line has an uncountable number of them—a result of the denumerability of the rational numbers and the non-denumerability of the irrationals. Yet not even the most powerful microscope can reveal these holes, since both the rationals and the irrationals are dense along the number line. This simple yet strange function was given at the end of the nineteenth century as an example that functions may exist that are *nowhere continuous*. The controversy that such examples aroused brought about an entire reexamination of the deceptively simple concept of continuity.

12 Inversion in a Circle

A function may be thought of as a transformation, or "mapping," from the x-axis to the y-axis, both of which are one-dimensional sets of points. In higher mathematics we also deal with transformations from a two-dimensional set of points to another two-dimensional set, that is, from one plane to another. One of the most interesting transformations of this kind is the transformation of *inversion*, or more precisely, *inversion in the unit circle*. Given a circle with center O and radius 1, a point P whose distance from O is $OP = r$ is "mapped" to a point Q, lying on the same ray from O as P, whose distance from O is $OQ = 1/r$ (Fig. 12.1). In this way, a one-to-one correspondence is established between the points of the original plane and those of the new plane: every point of the one plane is mapped onto a point of the other.[1] There is only one exception to this rule: the point O itself. To

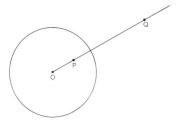

Figure 12.1. *Circular inversion.*

---

[1] Of course, we may think of the two planes as coinciding, in which case the mapping can be thought of as a movement of the points of this plane.

see this, let us follow the whereabouts of the "image" point Q as the point P moves in the plane. Our rule of correspondence requires that $OQ = 1/r = 1/OP$. Therefore, the closer P is to O (the smaller the value of r), the farther will Q be from O. As P approaches O, Q recedes to infinity. The point O, therefore, does not have any definite image, and we must exclude it from our transformation.[2] And yet it is tempting to define a "point," the *point at infinity,* which is the image of O under inversion. Of course, this "point" is not a point at all, at least not in the ordinary sense of the word: we cannot locate it anywhere in the plane, and it does not determine any definite position, as an ordinary point does. Its sole purpose is to enable us to carry out the transformation without any restrictions. It is in this sense, and in this sense only, that we may write the pair of equations

$$\frac{1}{0} = \infty, \qquad \frac{1}{\infty} = 0$$

Inversion has many interesting properties, all of which are related in one way or another to the point at infinity. First, all points inside the circle of inversion—including now the center O—are mapped onto points outside it, and vice versa; this is because if $r < 1$, then $1/r > 1$, and conversely. We may thus think of the interior of the unit circle as a condensed image, a microcosmos, of its exterior (Fig. 12.2). Of course, if we exclude from

Figure 12.2. *Inversion maps the interior of the unit circle onto its exterior, and vice versa.*

[2] This, of course, also follows from the fact that for the point O we have $r = 0$, making the division by r invalid.

our discussion the point at infinity, then we must leave a hole at the center of the unit circle, for the center is not mapped onto any ordinary (finite) point. It is precisely to avoid such a hole that the point at infinity has been introduced.

A second inherent feature of inversion is its *symmetry:* if Q is the image of P, then also P is the image of Q. This follows from the defining equation $OQ = 1/r = 1/OP$, which can be written as $OP = 1/(1/r) = 1/OQ$, showing that P is obtained from Q by the same rule as Q from P. Thus, the words "point" and "image" can always be interchanged, which means that every statement about inversion that holds for a given set of points also holds for the set of image points. This symmetry has a great unifying power: it enables us to make general statements about inversion without distinguishing between "point" and "image." It is the point at infinity which makes this symmetry possible, for without it we would have to exclude from every statement the center of inversion O.

Next we ask how a given curve will change when each of its points is subjected to inversion. We immediately see that the points of the unit circle (i.e., all points for which $r = 1$) are mapped onto themselves. The unit circle thus does not change at all—it remains *invariant* under the transformation. It is also easy to see that any straight line through the center O remains invariant as a whole, although an exchange of points along the line takes place: points of the line lying inside the unit circle are exchanged with points lying outside it, and vice versa. But much more surprising is the fact that straight lines *not* passing through O are mapped onto *circles that pass through O*, and conversely. (A proof of this fact is given in the Appendix.) Figure 12.3 shows three such lines: In (a) the line l is outside the circle of inversion c, and therefore its image (the circle k) lies entirely inside c. In (b) l touches c from the outside, hence k touches c from the inside. Finally in (c) l crosses c at the two points P and Q, resulting in an image circle k that passes *through* O, P, and Q (the portion of this circle lying inside c is the image of the outside part of l, and conversely). Thus, the closer a line is to the center of inversion O, the larger will its image circle be. Note that all three image circles pass through O; this is because every straight line passes through the point at infinity, and therefore its image must pass through the center of inversion, which is the image of the point at infinity. We will return to this issue shortly; meanwhile, let us enjoy the beautiful pattern shown in Fig. 12.4, which shows an ordinary checkerboard on which its own inversive image is superimposed. All the above-mentioned features are exhibited here; we clearly

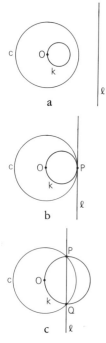

Figure 12.3. *Under inversion, a straight line not passing through the center O is mapped onto a circle passing through O, and vice versa. Three different positions of the line and its image circle are shown.*

90

Figure 12.4. *A checkerboard and its image under inversion. The small empty space near the center is the image of the entire plane outside the checkerboard limits. Reprinted from Harold R. Jacobs,* Geometry, *W.H. Freeman, 1974, with permission. Adapted from "Mathematical Games" by Martin Gardner.* Copyright © 1965 by Scientific American, Inc.

see how the circle of inversion turns the outside squares into the small circular regions inside it. The tiny empty space around the center is, of course, the image of the entire plane outside the limits of the board.

Not every curve, of course, has an image as simple as a straight line or a circle. In particular, the family of conic sections (except for the circle) is not preserved under inversion, but is transformed into curves whose mathematical description is quite complicated. Figure 12.5 shows how an ellipse, a parabola, and a hyperbola change under inversion. The small ellipse-looking figure in (a) is deceiving in its simplicity, for it is not an ellipse at all! (It turns out that the image of an ellipse is a curve represented by an equation of the fourth degree, while the ellipse itself is expressed by a second-degree equation.) The parabola turns into the graceful curve in (b), known as a *cardioid*—the cusp at the center is the image of the point at infinity, where the two prongs of the parabola meet. Finally, the hyperbola becomes the eight-shaped figure shown in (c), which resembles a tilted infinity symbol.

We see, then, that inversion greatly distorts the shape of some curves, while leaving others totally unchanged. This should not really come as a surprise, since inversion carries the "very near" to the "very far" and consequently must compress figures that stretch to infinity into figures that occupy only a finite region of the plane. We can gain a good insight into this process from Fig. 12.6, which shows how the individual points on a circle k through 0 are carried to their image points on the straight line

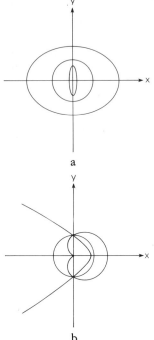

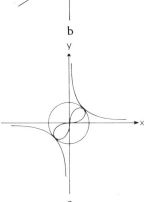

Figure 12.5. *The images of an ellipse, a parabola, and a hyperbola under inversion.*

a

b

c

Figure 12.6. *Details of the transformation of a straight line into a circle under inversion.*

l. As the point *P* travels around *k*, its image *Q* moves along *l*. As *P* approaches *O*, *Q* recedes to infinity along one direction of *l* and disappears there. But hold! As *P* passes *through O*, *Q* reemerges from the other direction of *l*. It seems, therefore, that there is a discontinuity in the way in which inversion carries a circle to its image line, for, depending on whether we approach *O* in a clockwise or counterclockwise direction, we can go to infinity along one or the other direction of the line. But this apparent discontinuity is removed once we define a *single* point at infinity (and not two such points, as we might be tempted to do)—the image of *O* under inversion. It is precisely this definition which

Figure 12.7. *M.C. Escher:* Hand with Reflecting Sphere (*1935*). © *M.C. Escher Heirs c/o Cordon Art—Baarn—Holland.*

is forced upon us if we want to preserve the one-to-one nature of our transformation and its inherent symmetry.

But let us return to the family of straight lines as a whole. We have seen that this family is transformed by inversion into either straight lines or circles. Now the symmetry of our transfor-

93

mation gives us a good reason to suspect that the converse of this statement is also true: that the family of all *circles* will be transformed into straight lines or circles. This indeed turns out to be the case, although the proof is somewhat less elementary than that of the previous fact. A bold idea therefore presents itself: If we agree to regard a straight line as a limiting case of a circle with an infinite radius, then all our conclusions can be summarized in one brief statement: *Inversion always transforms circles into circles,* where "circles" now includes straight lines. We see, then, that inversion really removes the distinction between these two curves, with their multitude of properties studied in school geometry. Straight line and circle—the classic antitheses of "straight" and "curved"—become united under it. And it is once again the point of infinity which makes this unification possible.

We may, of course, extend our discussion to three dimensions and define inversion in the unit *sphere.* Most of the properties we have discussed will still be valid, except that lines must be replaced by planes and circles by spheres. In this form, the transformation is known to us from the familiar Christmas decoration balls, which seem to reflect the entire world into a small, distorted, image. The Dutch artist Maurits C. Escher (1898–1972), of whom we will have more to say later, depicted the nature of spherical reflections in his work *Hand with Reflecting Sphere* (1935), which we reproduce here (Fig. 12.7).[3]

---

[3] It must be said that three-dimensional inversion is only an approximation to spherical reflection, but this approximation does exhibit the main features of this kind of reflection.

Geographic Maps and Infinity   13

It is, of course, not just for the sake of beauty that mathematicians study inversion, for the subject turns up in many different branches of science, sometimes quite unexpectedly. We will discuss here one such case—the role inversion plays in cartography, the science of map making.

As is known to everyone who has tried to press the peels of an orange against a table, it is generally impossible to flatten a curved surface onto a plane without severely distorting the surface. This is true regardless of how carefully one tries to do the job—some warping will always take place. Hence, in order to draw the surface of the earth (or of the globe, which is a scaled-down model of the earth) on a flat map, we must devise some method of *projection*—a scheme that will transform every point on the surface of the earth onto a corresponding point on the map. Cartographers use many different kinds of projections, each with its particular advantages but also its characteristic distortions: no single projection can faithfully reproduce *all* properties of the earth, such as the distance between two points, the direction from one point to another, or the area of a region.

One of the most widely used projections, known already to Hipparchus in the second century B.C., is the *stereographic projection*. We imagine the earth (thought of as a perfect sphere) to be placed on a flat sheet of paper, touching it at the *south pole S* ((Fig. 13.1). We now connect every point P on the surface of the earth by a straight line to the *north pole N,* and extend this line until it meets the paper at a point P'. P' is the image of P under the projection.

We can easily establish the main features of this projection. First the circles of latitude (the "parallels," as they are known

Figure 13.1. *The stereographic projection.*

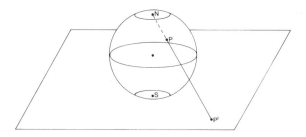

in geography), are mapped onto concentric circles around the south pole *S*, while the circles of longitude (the meridians) are mapped onto straight lines emanating from *S* like rays in all directions. (Fig. 13.2) The equator *E* on the earth goes over to the circle *e* on the map, which we may regard as the unit circle. The entire northern hemisphere is then mapped onto the exterior of *e*. The closer a point is to the north pole, the farther out will its image be on the map. There is one point on the earth without any image on the map: the north pole itself. Its image is at infinity.

These features immediately remind us of circular inversion, and indeed there is a close connection between inversion and the stereographic projection. Using trigonometry, it can be shown that two points on the earth having the same longitude but opposite latitudes are mapped onto two mutually inverse points on the map. (A proof of this and other properties of the projection is found in the Appendix.) The entire northern hemisphere on the

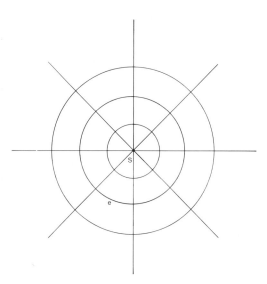

Figure 13.2. *Grid of the southern hemisphere under a stereographic projection.*

map may therefore be thought of as the inversive image of the southern hemisphere, and vice versa. One can show, furthermore, that every straight line or circle on the map corresponds to a circle on the earth. These features, in fact, often make it useful to carry out the projection in reverse—from the map to the earth. Used in this way, the sphere (now deprived of its geographical significance and looked upon as a purely geometric object) can be thought of as a finite "model" of the infinite plane, and this model preserves all the essential features of inversion. The advantage of such a model is that it makes it easier for us to visualize various aspects of planar inversion. Instead of inversion, we simply think of reflection in the equatorial plane of the sphere (i.e., a reversal of northern and southern latitudes); also, straight lines and circles always become circles, and the elusive point at infinity becomes a regular point on the sphere—the north pole. We will have another opportunity to look at such models later on in our discussion.

But let us return to geographical matters. Because of the increasing distortion that the stereographic projection suffers near the north pole, it becomes unsuitable for mapping the entire earth. In practice, only one hemisphere is mapped on any given map; of course, this could be the northern hemisphere, in which case we have to project the earth from the *south* pole. (Figure 13.3 shows how the northern hemisphere actually looks under this projection.) But to compensate for this shortcoming, our projection enjoys a property of great significance in cartography: it is direction-preserving, or *conformal.* By this we mean that if two curves on the earth (such as the paths taken by two ships at sea) intersect

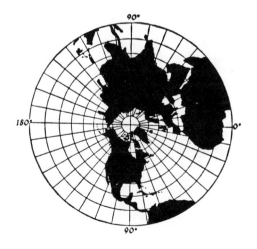

Figure 13.3. *The northern hemisphere under a stereographic projection. Reprinted from David Greenhood,* Mapping. *The University of Chicago Press, 1964, with permission (reproduced from Deetz and Adams).*

at a certain angle, their image curves on the map intersect at the same angle. The angle of intersection, in other words, remains invariant under the projection. It is this property that makes the stereographic projection an indispensable tool in navigation.[1]

Let us now imagine that we set out on a long journey on which we always follow a given, predetermined compass course—say 30° east of north. Let us decide to follow this course steadfastly, ignoring any obstacles we might encounter on our way, such as mountain ranges, seas, or impassable deserts. Such a course—one that keeps a constant direction relative to the north (and therefore crosses every meridian at the same fixed angle)—is known as a *loxodrome*. [2] For many years it had been thought that the loxodrome connecting two points on the earth is also the shortest distance between them, but this was disproved by the Portuguese Pedro Nunes (1502–1578). Nunes showed that while the shortest distance between two points is an arc of the great circle[3] joining them, the loxodrome is a spiral-like curve that winds around either pole indefinitely, approaching it ever closer but never reaching it. The Dutch artist Maurits C. Escher, whom we have already met in connection with spherical inversion, has depicted the nature of the loxodrome in his beautiful work *Sphere Spirals* (1958), shown in Fig. 13.4.

How will a loxodrome show up on a stereographic map? To answer this question, let us bring into use the conformal property

---

[1] Literally, the word "conformal" means "to have the same shape." We can easily see that the angle-preserving property of our projection also implies shape-preserving. Consider a small triangle on the earth. Each of its sides will show up as an arc segment on the map, which, because of the smallness of the triangle, can be regarded as a straight-line segment. Since our projection preserves angles, each pair of sides of the original triangle intersect at the same angle as their images on the map; the image triangle is thus similar in shape to the original (although not necessarily equal to it in size). Since any region can be regarded as the sum of many such triangles, the same conclusion holds for any small shape. It must be stressed, however, that this is true only locally, i.e., for small shapes, because for a large triangle the image sides will become increasingly curved. Theoretically speaking, the conformal property holds only for *infinitely small* shapes, but for practical purposes it does afford a good approximation even for continent-size shapes, as Fig. 13.3 clearly shows.
[2] The word comes from the Greek words *loxos* = slanted, and *dromos* = course, i.e., a slanted line. The name was proposed by the Dutch physicist Willebrord Snell (1591–1626), who is better known for a famous law in optics bearing his name. A lo.odrome is also known as a rhumb line.
[3] A *great circle* is a circle whose center coincides with the center of the sphere (examples are the equator and the meridians). As we will see later, great circles play the role of straight lines on a sphere.

Figure 13.4. *M.C. Escher: Sphere Spirals (1958).* © *M.C. Escher Heirs c/o Cordon Art—Baarn—Holland.*

of our projection. Since a loxodrome crosses every meridian at the same fixed angle, its image on the map must intersect every straight line through the center of the map at the same angle. But we have already met a curve having just this property: the logarithmic spiral. Hence, on a stereographic map of the northern hemisphere, any loxodrome will be shown as a logarithmic spiral emanating from the north pole and unwinding as one goes southwards. For the southern hemisphere the situation will be reversed: each spiral will converge towards the south pole, strangling it indefinitely but never reaching it.

Now the logarithmic spiral may have a great aesthetic appeal to a mathematician, but it is an awkward curve to plot on a map—one can use neither a straightedge nor a compass. It is for this reason that the Flemish geographer Gerhardus Mercator (also know as Gerhard Kremer, 1512–1594) devised a conformal map on which all loxodromes show up as straight lines. His celebrated map, published in 1569, shows the entire globe on a rectangular coordinate grid on which all parallels (lines of latitude) are of

99

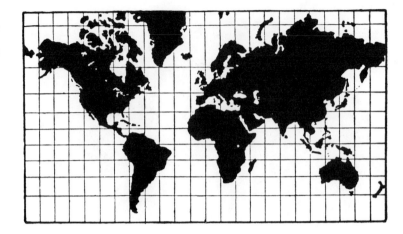

equal length, as are the meridians (Fig. 13.5). Now on the globe, the circles of latitude become smaller and smaller as one approaches the poles; on Mercator's map, therefore, each parallel is stretched beyond its correct length by a certain ratio that depends on the latitude of the parallel: the higher the latitude, the greater the stretching. Therefore, if the map is to be conformal, one must compensate for this stretching by an equal stretching of the spacings *between* the parallels. By working out this scheme, Mercator was able to calculate the correct spacing for each parallel.[4] The result was a map on which every loxodrome crosses every meridian at a constant angle, and since all the meridians are parallel, all loxodromes show up as straight lines.

---

[4] Contrary to erroneous statements found in many geography books, Mercator's projection is not obtained by wrapping the globe with a cylinder around the equator, and then projecting from the center of the globe onto this cylinder. Strictly speaking, Mercator's projection is not a projection at all: it can only be obtained by a mathematical formula whose derivation requires the knowledge of calculus. Mercator, who lived a full century before the invention of the calculus and therefore could not have benefited from its advantages, found the necessary spacings by a process known today as "numerical integration"—the solution of a differential equation by a step-by-steep process, starting with an initial value and then proceeding gradually from point to point. This process is efficiently executed on a modern computer, or even on a hand-held calculator, a luxury which again Mercator could not enjoy. This makes his achievement all the more remarkable, and his map is regarded by many as the single greatest contribution to the science of cartography since the discovery that the earth is round. For more details, see *Elements of Map Projection* by Charles H. Deetz and Oscar S. Adams, Greenwood Press, New York, 1969.

Mercator's map had an immediate and decisive impact on the nautical community of his day, for it enormously simplified the art of navigation. No more would a mariner have to struggle with awkward curves on his map; from then on, all he had to do was to plot his intended course as a straight line connecting his points of departure and destination, measure the direction of this line relative to the north, and then follow the same direction at sea. But as is always the case in cartography (and elsewhere as well), the attainment of one goal comes at the expense of another, in this case the shape of the earth. Because of the increased stretching of the distance between the parallels as we approach the poles, countries in high latitudes appear greatly distorted on the map; the island of Greenland, for example, appears larger than South America, although in reality it is nine times smaller. It is this north–south stretching that gives Mercator's map its characteristic look. Of course, neither the north nor the south pole can be shown on this map, for their images are at infinity.

14 Tiling the Plane

Threefold is the form of space:
Length, with every restless motion,
Seeks eternity's wide ocean;
Breadth with boundless sway extends;
Depth to unknown realms descends.

— Friedrich von Schiller (1759–1805)

But let us return to ordinary geometry. Among the host of geometric figures around us, the regular polygons have always played a special role. A *polygon* (from the Greek words *polys* = many and *gonon* = angle) is a closed planar figure made up of straight line segments. A *regular* polygon is a polygon whose sides and angles are all equal. The simplest regular polygon is the equilateral triangle; next comes the square, followed by the pentagon, the hexagon, and so on. As we saw in Chapter 1, the Greeks were particulary interested in these regular polygons and used them to find an approximation for the number π. They knew, of course, that there exist infinitely many of these polygons; that is, for any given integer $n \geq 3$, there exists a regular polygon having n sides—an "n-gon", as mathematicians say.

A fundamental question arising in connection with regular polygons is: Which of them can tile a flat surface, such as a floor? By "tiling" (the word "tessellation" is also used) we mean that we can fill the entire plane with the endless repetitions of a single basic design, without leaving any empty spaces between. It is not hard to show that among the infinity of regular polygons, only three will do the job: the equilateral triangle, the square, and the hexagon.[1] We can easily see why a regular pentagon, for example, will not tile the plane: two adjacent sides of the pentagon form an angle of 108° (Fig. 14.1); now for any number of regular

---

[1] We do not distinguish, of course, between regular polygons of the same shape but different size; that is, if a basic figure will tile the plane, so will an enlargement or a contraction of the same figure.

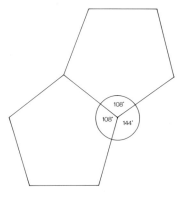

Figure 14.1. *A regular pentagon cannot tile the plane.*

pentagons to meet at a vertex and fit tightly, 108° would have to go an integral number of times into a full rotation, that is, into 360°—which is not the case. In the same way, all other *n*-gons can be eliminated except the 3, 4, and 6-gons. We are all familiar, of course, with tiling by a square—the most obvious form of tiling, since it will not only cover the entire floor, but also fit exactly against the walls of a rectangular room. Tiling with hexagons can be found in many promenades—the subway stations of Washington, D.C., provide a good example. Tiling by equilateral triangles is essentially equivalent to hexagonal tiling, since six such triangles make up a regular hexagon. Hexagonal tiling can also be found in nature—it forms the basic structure of a honeycomb. There is a reason for this, which apparently must be known to the bees: of the three tilings, the one by hexagons is the most efficient—it requires the least amount of material to cover a given area. The three basic tilings are shown in Fig. 14.2.

The situation is entirely different when we move up from two to three dimensions. Here a surprise is waiting for us: There are exactly five different regular solids, and of these, only one can "tile" space. By a regular solid, or *polyhedron,* we mean a solid whose faces are all congruent regular polygons that meet each other at the same angle in space. The fact that there exist just five regular solids—unlike the infinitely many regular plane polygons—is a result of a famous formula discovered in 1752 by Leonhard Euler: For any simple polyhedron (a polyhedron having no holes), the number of faces *F,* the number of edges *E,* and the number of vertices *V* are always connected by the equation

$$V - E + F = 2$$

From this formula one can show (see the Appendix) that there

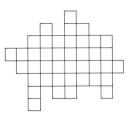

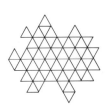

Figure 14.2. *The three possible tessellations by regular polygons.*

103

are only five possibilities of obtaining a regular polyhedron:

$$V = 4, \quad E = 6, \quad F = 4: \quad \text{the tetrahedron}$$
$$V = 8, \quad E = 12, \quad F = 6: \quad \text{the cube}$$
$$V = 6, \quad E = 12, \quad F = 8: \quad \text{the octahedron}$$
$$V = 20, \quad E = 30, \quad F = 12: \quad \text{the dodecahedron}$$
$$V = 12, \quad E = 30, \quad F = 20: \quad \text{the icosahedron}$$

(Except for the cube, the names are derived from the number of faces.) These regular polyhedra, also known as the Platonic solids (Fig. 14.3), were well known to the Greeks. Many mythical

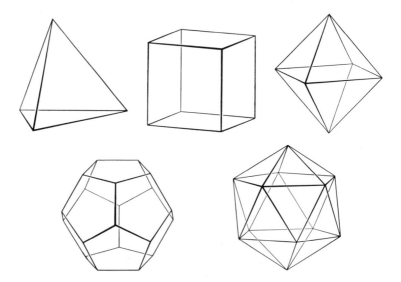

Figure 14.3. *The five regular or Platonic solids. Reprinted from D. Hilbert and S. Cohn-Vossen,* Geometry and the Imagination, *Chelsea Publishing Company, 1952, with permission.*

Figure 14.4. *The icosahedron on the emblem of the Mathematical Association of America. This regular solid has twenty identical faces, each being an equilateral triangle. Courtesy of the Mathematical Association of America.*

attributes were ascribed to them, and the astronomer Johannes Kepler (1571–1630) gave them a prominent role in his theory of planetary orbits (a role which they did not deserve, as it turned out). In our time these solids have enjoyed a comeback in crystallography. The atoms of crystals are arranged in lattices which exhibit a strict spatial regularity. We may think of the lattice as being made up of "unit cells"—basic building blocks which, by their endless repetitions, fill up the volume of the crystal. The atoms of a diamond, for example, occupy the vertices of regular tetrahedra, and it is this arrangement that gives a diamond its celebrated hardness. A regular icosahedron appears on the emblem of the Mathematical Association of America (Fig. 14.4).

Euler's formula gives rise to a remarkable property of the five regular solids: if we connect by straight lines the centers of all faces of a given regular solid, we again get a regular solid, called the *dual* of the original solid. Thus, the dual of a cube is an octahedron, and vice versa (Fig. 14.5); that of an icosahedron is a dodecahedron, and vice versa. The tetrahedron is self-dual: its dual is again a tetrahedron, albeit a smaller one. This property is a result of the fact that the variables F and V in Euler's formula appear in a symmetric way: we can interchange them without affecting the correctness of the formula. (Note that this is not true for E, which appears with a negative sign.)

It is this property which immediately gives rise to an infinite progression of dual regular solids. We can start, for example, with a cube, connect the centers of its six faces and get a regular octahedron, then do the same with its eight faces and get once again a cube, and so on *ad infinitum*. The progression of solids thus obtained will diminish in size, each solid being nested within its predecessor. The dimensions of the solids in this progression follow exact numerical ratios; except for the tetrahedron, however, these ratios are far from simple. For the tetrahedron, which is self-dual, the ratio is $1/3$: Each tetrahedron has a side of length one-third that of its predecessor (and therefore a volume 27 times as small). For the cube–octahedron progression, the ratio is $\sqrt{2}/2$ when going from a cube to an octahedron, and $\sqrt{2}/3$ when going from an octahedron to a cube. For the icosahedron–dodecahedron sequence, the ratios are much more complex.

Returning once more to the plane, we have seen that there are only three regular polygons which will tile the plane. But if we relax some of the restrictions on the polygons, many new possibilities arise. Thus, it is possible to tile the plane with a pentagon of equal sides but unequal angles, as shown in Fig. 14.6. (Note that there are two pairs of equal angles in each pentagon.) As the figure shows, four such pentagons can always be combined to form a hexagon, so that the pentagon tessellation is at the

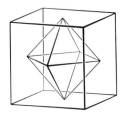

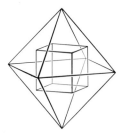

Figure 14.5. *Duality: the dual of the cube is the octahedron, and vice versa. Reprinted from D. Hilbert and S. Cohn-Vossen,* Geometry and the Imagination, *Chelsea Publishing Company, 1952, with permission.*

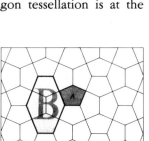

Figure 14.6. *The shaded pentagon A, consisting of equal sides but unequal angles, tiles the plane. Note that four such pentagons combine to form a hexagon. From Phares G. O'Daffer and Stanley R. Clemens,* Geometry: An Investigative Approach. *Copyright © 1976 by Addison-Wesley Publishing Company, Inc. Reprinted by permission.*

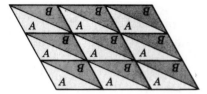

Figure 14.7

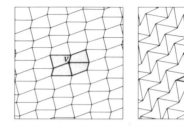

Figure 14.8

Figure 14.7. *Any triangle will tile the plane. From Phares G. O'Daffer and Stanley R. Clemens,* Geometry: An Investigative Approach. *Copyright © 1976 by Addison-Wesley Publishing Company, Inc. Reprinted by permission.*

Figure 14.8. *Any quadrilateral (four-sided polygon) will tile the plane. From Phares G. O'Daffer and Stanley R. Clemens,* Geometry: An Investigative Approach. *Copyright © 1976 by Addison-Wesley Publishing Company, Inc. Reprinted by permission.*

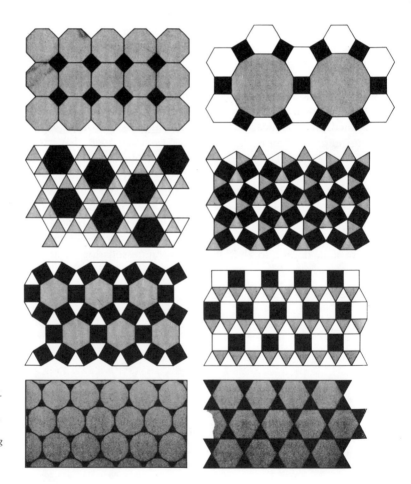

Figure 14.9. *The eight possible semi-regular tessellations. From Phares G. O'Daffer and Stanley R. Clemens,* Geometry: An Investigative Approach. *Copyright © 1976 by Addison-Wesley Publishing Comapny, Inc. Reprinted by permission.*

106

same time also a hexagon tessellation. (Note that this is not the regular hexagon of the honeycomb tessellation, but an elongated one, with four long and two short equal sides.) It is known that no tessellation is possible with any polygon of more than six sides. On the other hand, it is easy to demonstrate that *any* triangle and *any* quadrilateral (4-gon) will tessellate a plane, as shown in Figs. 14.7 and 14.8.

There is also the possibility of tiling a plane with more than one basic polygon. A careful analysis shows that there are exactly eight combinations of regular polygons which tile the plane in such a way that every vertex is surrounded by the same type and number of regular polygons.[2] These eight combinations, known as the *semi-regular* tessellations, are shown in Fig. 14.9. Finally, we may drop even the requirement that all vertices should be identical, giving rise to tessellations of remarkable complexity, such as the one shown in Fig. 14.10. The past two decades have seen an enormous surge of interest in this subject,[3] inspired, no doubt, by the work of the Dutch artist Maurits C. Escher, who in turn got his inspiration from the ornamental art of the Moors. Indeed, it is in the visual arts that the idea of tessellation has found its greatest application. We will turn our attention to these matters in Part III.

Figure 14.10. *A more complex tessellation. From Phares G. O'Daffer and Stanley R. Clemens,* Geometry: An Investigative Approach. *Copyright © 1976 by Addison-Wesley Publishing Company, Inc. Reprinted by permission.*

Figure 14.11. *Except for the cube, none of the five regular solids can fill space. But the truncated octahedron shown here, know already to Archimedes and rediscovered by the Russian crystallographer Fedorov, can indeed fill up three-dimensional space without leaving any gaps. From D'Arcy W. Thompson,* On Growth and Form, *edited by J.T. Bonner. Copyright © 1961 by Cambridge University Press. Reprinted by permission*

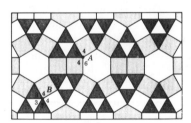

Figure 14.10

Figure 14.11

[2] The details are rather technical and can be found in *Geometry: An Investigative Approach,* by Phares G. O'Daffer and Stanley R. Clemens, Addison-Wesley, Menlo Park, California, 1977. This book also has an extensive bibliography of articles related to tessellation.

[3] See, for example, the articles "Pentaplexity: A Class of Non-Periodic Tilings of the Plane" by R. Penrose, *The Mathematical Intelligencer,* 2 (1979), pp. 32–37, and "Extraordinary Nonperiodic Tiling that Enriches the Theory of Tiles" by Martin Gardner, *Scientific American,* January 1977, pp. 110–121.

15 A New Look at Geometry

All finite things reveal infinitude.

— Theodore Roethke (1908–1963),
The Far Field, IV

We close Part II with an examination of two of the most revolutionary developments in modern mathematics—both directly related to infinity. The first of these, the creation of projective geometry, takes us back to the Renaissance, and it has its roots not in science but in art. During the Middle Ages, both science and art were subordinated to the religious and mythological beliefs of the time. Nature was depicted not as she really was, but as the observer's fantasies and religious beliefs wanted her to be. Thus, the world believed in a sun that moved around the earth, not because the available evidence, based on an objective observation of the heavens, made such a conclusion inevitable, but because the Roman Catholic church decreed that it must be so. The earth itself was flat—despite mounting evidence to the contrary—because to believe in a round earth meant to let the poor creatures on the "other side" plunge into the abyss of infinite space. And a painter depicted his saints and heroes not in their natural perspective— that is, faraway figures appearing smaller than nearby ones—but according to their status in the Church hierarchy.

It was not until the fifteenth century that a more objective observation of nature began to take place; and once the winds of change began to blow, they could not be stopped. A true, faithful description of our world became the motto of the new era, and it left scarcely a facet of Western society unchanged. In art, particularly, the new outlook resulted in an attempt to study the laws by which our eyes perceive the sights and objects around us. Thus was born the art of perspective, founded in 1425 by the Italian architect and sculptor Filippo Brunelleschi (1377–1446) and brought to perfection by Albrecht Dürer and Leonardo da Vinci.

Perspective is a graphical scheme which enables the artist to put on canvas the scene in front of him in a true and objective manner. A picture drawn according to the laws of perspective is a sort of photograph—a faithful reproduction of what the eyes actually see. But we know that a photograph of a scene differs from the actual scene in several ways: a circle, for example, may appear as an ellipse, a square might show up as a trapezoid, and a pair of parallel lines—such as railroad tracks—seem to converge on the horizon. It was precisely this question—how does an object appear when depicted on canvas—that in the sixteenth century gave rise to an entirely new branch of mathematics, projective geometry.

Unlike classical Greek geometry, which is exclusively concerned with the size and shape of figures, projective geometry deals with a more fundamental, and in a way simpler, aspect of figures: those properties which remain unchanged under a *projection.* By "projection" we mean the collection of all rays of light emanating from the object and converging at the eye (or the camera's lens, as the case may be). Now when an artist depicts a scene on his canvas, we can think of the canvas as a portion of a plane, or *section,* that intersects these rays of light. The picture is then formed by the collection of all the points where the projecting rays pierce the canvas. Thus the basic goal of projective geometry is to study the relations between an object and its image on the section, and in particular those properties of the image which remain unchanged, or invariant, under the projection. It is here that the notion of infinity enters our discussion.

We have already seen that two parallel lines do not, in general, appear to be parallel on a picture. Thus the property of parallelism is not preserved under a projection. But we also know from experience that parallel lines *seem* to converge on the horizon. Now what kind of an object is this horizon, where lines that do not meet appear to meet after all? Projective geometry regards it as an ordinary line, but a special one nonetheless: the "line at infinity." Every pair of parallel lines in the plane defines a point, the "point at infinity," where the two lines seem to converge; and the totality of all points at infinity constitutes the line at infinity, the horizon (Fig. 15.1).[1] Thus projective geometry formally endorses the popular—and vague—phrase, "parallel lines meet at infinity."

---

[1] The names "ideal point" and "vanishing point" are also used, and likewise "ideal line" and "vanishing line."

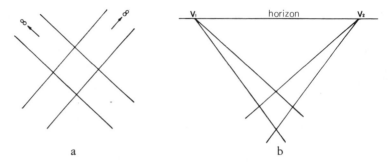

Figure 15.1. *Two pairs of parallel lines as seen: (a) from above; (b) in perspective.*

At first thought, the introduction of a vague concept like this into geometry may well defy our image of mathematics as an exact discipline, a science in which there is no room for vagueness and ambiguity. True enough! Our new creation will become legitimate only if we can give its concepts a precise definition, and then justify this definition by its usefulness. As for the definition, we have just given it when we agreed that two parallel lines meet at a point at infinity. Note that this in no way contradicts the traditional definition of parallel lines as lines (in one plane) that do not meet, for classical geometry never deals with infinity, only with finite points; thus the traditional definition only says that parallel lines do not meet at any *finite* point. The addition of the points and line at infinity, therefore, merely replaces this negative statement (i.e., what parallel lines do *not* do) with a positive one.

Now we can, of course, define as many new concepts as we please, but the ultimate justification of doing so rests on whether these concepts shed some new light on the subject under consideration. It is here that the advantage of our new outlook becomes evident, for the introduction of the points and line at infinity has a great unifying power: it eliminates the need to treat parallel lines differently from "ordinary" lines. In ordinary geometry we say, "two intersecting lines meet at one and only one point; parallel lines do not meet." In projective geometry we say instead, "any two lines meet at one and only one point." There is a certain elegance in such a statement, with its sweeping generality; and mathematicians, like artists, always look for elegance in their work.

There is, to be sure, quite a difference between ordinary points and points at infinity. While an ordinary point determines a *position,* a point at infinity determines a *direction.* The point of intersection of two non-parallel lines gives us the exact position where the two lines meet; the point of intersection of two parallel lines merely tells us their direction. And since all lines parallel to one another have the same direction, we will agree further that all such lines shall meet at a single point at infinity. That is, every family of parallel lines has its own unique point at infinity.[2]

As we have just seen, the introduction of the points and line at infinity makes it unnecessary to distinguish between intersecting and parallel lines. Were it only for this achievement, however, projective geometry could hardly justify itself as a separate branch of mathematics. As it turns out, our new outlook actually enables us to discover a host of new results which would have remained totally unnoticed had we used only the methods of classical geometry. Let us examine, for example, the following two simple statements, both taken from elementary geometry:

Two points determine one and only one line.
Two lines determine one and only one point.

(See Fig. 15.2; the second statement, of course, includes the case where the two lines are parallel.) What strikes one about these two statements is their complete symmetry: the words "points"

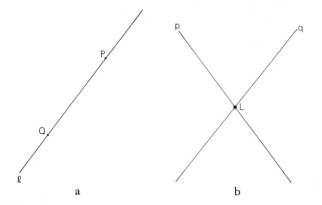

a   b

Figure 15.2. *Duality:* (*a*) *one line passes through two points;* (*b*) *one point lies on two lines.*

---

[2] Contrary to the intuitive notion that a line stretches to infinity in both directions, we adopt here the viewpoint that an ordinary line has just one point at infinity. This is done in order to preserve the axiom that through two given points, one and only line passes. Had we assigned *two* points at infinity for each line, then we would have been forced to regard all parallel lines as identical.

Note that this convention differs from the one we adopted for inversion, where we defined a single point at infinity for the entire plane. Accordingly, one distinguished between the ordinary or *Euclidean plane,* which has no points at infinity, the *inversive plane* (one point at infinity), and the *projective plane* (a point at infinity for each family of parallel lines).

These ideas can easily be extended to three dimensions: Every family of parallel *planes* will meet at line at infinity, and since there are infinitely many such lines (corresponding to the infinitely many families of parallel planes in three-dimensional space), we define a single *plane at infinity* in which all these lines lie.

111

and "lines" appear in them in exactly the same way. In fact, we could interchange these two words without affecting the validity of either statement; the only change would be that the first statement would become the second, and vice versa. This complete symmetry is known as the *principle of duality,* and it is one of the most beautiful achievements of projective geometry.

According to the principle of duality, any valid statement about the mutual relationship of points and lines—a statement of incidence, as it is called—will retain its validity if the words "point" and "line" are everywhere interchanged. We say that the two statements, and their corresponding geometric configurations, are "dually equivalent." For example, a triangle may be thought of as either a set of three non-collinear points (points not all of which lie on one line), as in Fig. 15.3a, or as a set of three non-concurrent lines (lines not all of which pass through one point), as in Fig. 15.3b. The first definition is the more common one, but the second is just as valid. (Of course, we then have to think of the lines as being extended indefinitely, which gives our triangle a somewhat unusual appearance.) This result, however, is rather trivial. A much more interesting fact, discovered in 1639 by the French architect and engineer Gérard Desargues (*ca.* 1593–1662) and named after him, says:

> If two triangles ABC and A'B'C' are so placed that the lines AA', BB', and CC' pass through one point, then the points aa', bb', and cc' lie on one line.

Here we denote by AA' the line passing through the points A and A', and by aa' the point "passing" through the lines a and a' (that is, the point of intersection of these two lines); likewise

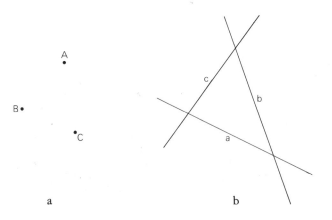

Figure 15.3. *Dual interpretation of a triangle: (a) three non-collinear points; (b) three non-current lines.*

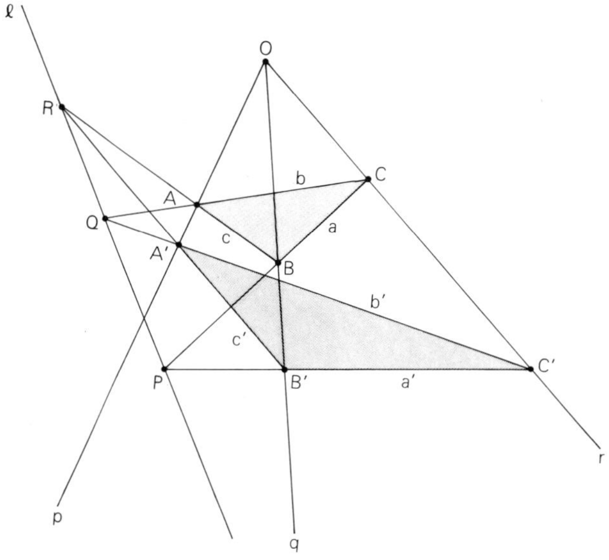

Figure 15.4. *Desargues's theorem: the general case.*

for the other pairs of points and lines (Fig. 15.4). This choice of notation is entirely in the spirit of projective geometry, which strives towards a complete symmetry between points and lines.

Desargues's theorem is not at all easy to prove with the tools of ordinary geometry. But when viewed in the light of projective geometry, the proof becomes extremely simple. The two triangles *ABC* and *A'B'C'* can be thought of as projections of each other, where the point from which the projection is being made is the point *O* at which the lines *AA'*, *BB'*, and *CC'* meet. Note that for this to happen, the two triangles need not at all be in one plane; on the contrary, think of them as belonging to two planes inclined to each other. Now two inclined planes always meet in a straight line, and therefore the points of intersection of the corresponding sides must lie on this line. This completes the proof. Not only is this proof a model of simplicity, but it also provides an example where a statement about a two-dimensional geometric configuration is easier to prove when viewed from a three-dimensional point of view.

One may ask, what happens to Fig. 15.4 if the lines *AA'*, *BB'*, and *CC'* are parallel?[3] Will we have to amend Desargues's theorem for this special case? The answer is, not at all! If the three lines

---

[3] This is known as *parallel projection,* as opposed to the *central projection* we have seen before, in which the center of projection is an ordinary (finite) point.

Figure 15.5. *Desargues's theorem: the parallel case.*

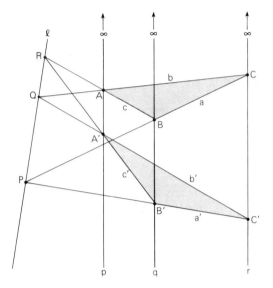

are parallel, this merely implies that their common point of inter-section *0* recedes to infinity, and projective geometry regards all points at infinity as bona fide points. Thus the points *aa'*, *bb'*, and *cc'* will still lie on one line (Fig. 15.5). Another case of interest arises when the two triangles are similar and lined up, as in Fig. 15.6. (We may think of them as two floors in a pyramid with a triangular cross section.) Then the six sides are parallel in pairs,

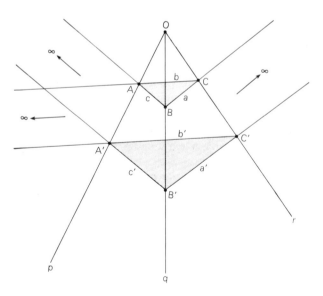

Figure 15.6. *Desargues's theorem: the case of similar triangles.*

with side *a* parallel to side *a'*, *b* to *b'*, and *c* to *c'*. The points
of intersection of these pairs will therefore recede to infinity, where
they lie on one line—the line at infinity! No matter how hard
we try to find a situation in which Desargues's theorem breaks
down, we fail; the points and line at infinity guarantee the validity
of the theorem in all cases.

Now what happens if we "dualize" Desargues's theorem? According to the principle of duality, this should give us another
valid theorem:

> *If two triangles ABC and A'B'C' are so placed that the points aa', bb',
> and cc' lie on one line, then the lines AA', BB', and CC' pass through
> one point.*

But this new theorem is precisely the *converse* of the original one.
That is, by merely interchanging the words "point" and "line"
in Desargues's theorem, we have obtained a new theorem in which
the premises and the conclusions of the original theorem are reversed. It happens very rarely in ordinary geometry that a theorem
and its converse can be obtained—and proved—from each other
by a mere change of words, but this indeed is the case here.

There is one other aspect of Desargues's theorem which is worth
mentioning. If we look again at Fig. 15.4, we see that it involves
ten points (the points O, A, B, C, A', B', C', P, Q, and R, where
the last three stand for *aa'*, *bb'*, and *cc'*, respectively) and ten
lines (the lines *l*, *a*, *b*, *c*, *a'*, *b'*, *c'*, *p*, *q*, and *r*, where the last
three stand for *AA'*, *BB'*, and *CC'*, respectively), so that the number of points and lines is equal. Now this, of course, is not much
of a surprise, in light of the principle of duality. But what *is* surprising is that all ten points and all ten lines are exactly equivalent
to each other: every point could play the role of every other
point, and similarly for the lines. For instance, we could regard
R as the point from which the projection is made, so that triangle
QAA' is projected on triangle PBB'. Then the points O (= *pq*),
C (= *ab*), and C' (= *a'b'*) all lie on one line, the line *r*. Since
in Desargues's configuration three lines pass through each point
and three points lie on each line, there are altogether 120 different
ways of arranging these points and lines, and for all of them the
original statement—and its converse—will be valid![4]

Projective geometry abounds with beautiful theorems about
points and lines, and in many older geometry books one could

---

[4] As a side bonus, Desargues's theorem provided the solution to an old
puzzle: how to plant ten trees in ten rows, three trees in each row. One
may reflect that in order to do this, one must know something about
what goes on very far away—at infinity!

Figure 15.7. *Dual interpretation of a curve: (a) as a set of points on the curve; (b) as a set of lines tangent to the curve.*

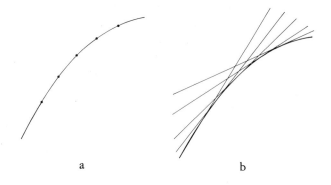

a   b

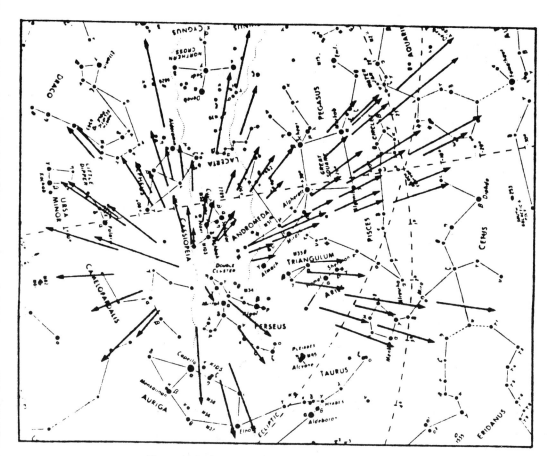

Figure 15.8. *A meteor shower: the meteors seem to emanate from one point in the sky, even though their paths are parallel. This point, known as the radiant, is a vanishing point—a point at infinity. Reprinted from* Sky & Telescope. *December 1980, with permission of Sky Publishing Corporation. Illustration by Ken Hodonsky.*

find pages neatly divided by vertical lines, with a statement and its dual appearing side by side across the line. But projective geometry did more than just introduce a measure of elegance into mathematics: it removed the point from the lofty position it had occupied in geometry since Euclid. In his great work, the *Elements,* written in Alexandria in the third century B.C., Euclid compiled into thirteen books the entire body of geometric knowledge as it was then known. This monumental work, which encompasses not only most of the geometry we learn in school but also topics from solid geometry and number theory, was the rock foundation of geometry for the next two millenia.[5] The very opening sentence of the *Elements* introduces the definition of a point: *"A point is that which has no part."* Next, after defining a line as a *"breadthless length,"* comes the definition of a straight line: *"A straight line is a line which lies evenly with the points on itself."* Thus the definition of a straight line already depends on the definition of a point. The notion that a line is made up of points had been so deeply rooted in our geometric intuition that no one seemed to have ever questioned it. Projective geometry, through the principle of duality, dispelled this notion: it put point and line on an equal footing, so that each could be regarded as *the* fundamental building block from which the rest of geometry is built.[6] Thus, by breaking with tradition, projective geometry freed itself from the dictums of the Greek classicists, opening the door for a flux of new discoveries which greatly enriched mathematics and affected its subsequent course. But all this could not have happened had we not introduced, right from the beginning, the points and line at infinity, for it is these two elements which enabled projective geometry to achieve its unifying goal.

---

[5] Howard Eves, in his *Introduction to the History of Mathematics,* has this to say about the *Elements:* "No work, except the Bible, has been more widely used, edited, or studied, and probably no work has exercised a greater influence on scientific thinking." We will have more to say about this work in the next chapter.

[6] An example of such a dual interpretation is shown in Fig. 15.7, in which a curve is regarded as a set of points lying on it (a) or as a set of lines tangent to it (b).

117

16 The Vain Search for Absolute Truth

As lines, so loves oblique, may well
Themselves in every angle greet
But ours, so truly parallel,
Though infinite, can never meet.

— Andrew Marvell (1621–1678)
from *The Definition of Love*

If projective geometry, through its principle of duality, has enormously enriched mathematics from an aesthetic point of view, the creation of non-Euclidean geometry has had an unparalleled intellectual impact on our entire scientific and philosophical thought. It marked the first serious doubt since Euclid's time as to the validity of our fundamental mathematical premises; it shattered the age-old belief in the power of mathematics to show us the road to the ultimate and absolute truth; and it brought about a whole reexamination of our ability to understand the physical world in which we live. This reexamination has had consequences which far transcended mathematics; ultimately, it led to the theory of relativity and helped in shaping our modern views of the universe. The spark that ignited this intellectual revolution was once again the infinite; more specifically, the question: What happens to parallel lines very far away?

The entire edifice of Euclidean geometry rests on a set of ten postulates, or axioms, whose truth Euclid regarded as self-evident, as so clear and unquestionable as to obviate the need for proof.[1] The fifth of these postulates says:

---

[1] These ten postulates are divided in the *Elements* into two groups: the first five deal with geometric concepts and are called "postulates"; the remaining five are arithmetic statements and are called "common notions." It is not entirely clear why Euclid made this distinction; in any case, we will follow the modern practice of regarding all ten statements as postulates, or axioms.

118

If a straight line falling on two straight lines makes the interior angles on the same side less than two right angles, the two straight lines, if produced indefinitely, meet on that side on which the angles are less than two right angles.

It is this axiom, which came to be known as the Parallel Postulate (or simply the Fifth Postulate), that is at the heart of the issue we are about to explore. Essentially, it sets the conditions for two straight lines in the same plane to intersect and, by implication, the conditions for two straight lines to be parallel (Fig. 16.1).

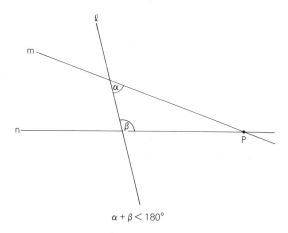

$\alpha + \beta < 180°$

Figure 16.1. *The Parallel Postulate: Euclid's version.*

Now the original formulation of the axiom, as stated above, differs considerably from the more familiar version we learn in school:

Given a line l and a point P not on l, there exists one and only one line m, in the plane of P and l, which is parallel to l.

This version (Fig. 16.2) is due to the Scottish mathematician John Playfair (1748–1819), who showed it to be equivalent to Euclid's version. But even though the two versions are equivalent (that is, each can be derived from the other), there is a slight difference in their appeal to our intuition: while Euclid's version never mentions the notion of parallelism—it merely sets the conditions for two straight lines to meet—Playfair's version explicitly talks about

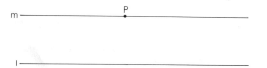

Figure 16.2. *The Parallel Postulate: Playfair's version.*

119

parallel lines, that is lines in the same plane that do not meet.[2]

In the course of the twenty centuries since Euclid's time, the Parallel Postulate has assumed something of a legendary status. Generation after generation of mathematicians have struggled with it, translating it into numerous equivalent forms and hoping, perhaps, to derive it from more fundamental principles. As we shall see, all these attempts have failed. But why was this axiom singled out from the other nine? Why did it receive so much attention? To answer this question, we must take a closer look at the logical structure of geometry and of mathematics in general.

The entire structure of classical geometry, with its numerous theorems about the various figures and shapes around us, rests on the ten axioms that open Euclid's *Elements*. These axioms have come to be regarded as the rock foundation of all of mathematics— a system of absolute truths whose validity is irrefutable. And if evidence for their validity should ever be sought, it is amply furnished by the physical world in which we live. Experience tells us, for example, that we can connect two points by one and only one line (a taut string, for example), which is the essence of Euclid's first axiom. It seems likewise obvious that a straight line can be extended indefinitely in both directions (the second axiom), or that all right angles are equal to one another (the fourth axiom). We must remember that Greek geometry was not an abstract science, created somewhere in the minds of scholars; on the contrary—it derived its concepts directly from the material world around us. Thus a point was regarded as an idealized version of a dot created by a pencil on paper, a line was an ideal straightedge extended indefinitely, and a right angle was the angle between a plumb line and the floor. It is not surprising, then, that the Greeks—and all their followers up until the eighteenth century— took the validity of Euclid's axioms for granted.

And yet there was a slight blemish. Somehow, some mathematicians felt that the Parallel Postulate was not quite as self-evident as the rest of the ten axioms. The reason lay with the definition of parallel lines: line which, *even if extended indefinitely,* will not meet. But how can we be sure that something will, or will not, happen very far away—indefinitely far, indeed? It is true that our intuition, supported by our daily experience, does indeed suggest that such should be the case. But then, the very philosophy

---

[2] The clause "lines in the same plane" is necessary, because two lines in space may cross each other without ever meeting and yet be non-parallel, as in the case of a road passing under another road. Such lines are known as skew lines.

120

of mathematics rests upon the need to provide a proof for any assertion whose truth is not immediately obvious. This means that every such assertion must be derived, by a purely logical deduction (i.e., by an argument free of any "physical" considerations), from other assertions which have already been proved before. Of course, we must begin somewhere, or else the process of mathematical proof will become an infinite regress. We must, in other words, decide what can be taken for granted, and what cannot. Assertions of the former type are *axioms,* or *postulates;* those of the latter type are *theorems,* or *propositions.* Ever since the sixth century B.C., when Greek mathematics began to evolve as an axiomatic science, this principle has been strictly and dogmatically followed: we agree on a set of self-evident assumptions, and from these we develop, by a chain of logical arguments, the entire structure. It is this principle which gives mathematics its character as a deductive discipline.

Of course, the strength of a theory rests not on its assumptions, but rather on its predictions. A theory that assumes too much at the outset is not likely to gain widespread respect among scientists. It is therefore incumbent upon a mathematician to reduce the number of axioms in a particular theory to a bare minimum, eliminating all the redundant ones and leaving only those which are absolutely necessary. The redundant axioms are those that can be derived from other axioms and must therefore be regarded as theorems. A true axiom, on the other hand, is logically independent of all the other axioms: it can neither be proved nor disproved.

Towards the end of the seventeenth century, several mathematicians began to suspect that the Parallel Postulate might, in fact, be redundant. It was not that anyone really doubted the validity of the postulate; rather, it was the feeling that it could perhaps be derived from the other nine axioms. Should this turn out to be the case, then the number of axioms could be reduced, and the logical structure and aesthetic appeal of geometry greatly enhanced. In the following centuries, numerous attempts would be made to prove the Parallel Postulate in one or another of its many equivalent versions. All would fail.

One of these attempts is particularly worth mentioning. The Italian Girolamo Saccheri (1667–1733), a Jesuit priest, decided to attack the problem by the so-called "indirect" method of proof. To prove a given proposition, we temporarily assume that the proposition is false, and therefore its negation is true. We then proceed to show that this assumption leads to a contradiction; therefore the negation is false, and the original proposition must

121

be true. Mathematicians, it must be said, are not generally too fond of this kind of proof. Whenever possible, they prefer a direct proof, the kind of proof that affirmatively establishes the truth of the proposition. There are situations, however, in which a direct proof is very difficult to arrive at; it is in such cases that we must resort to the indirect method, as Saccheri did.

Recall Mayfair's version of the Parallel Postulate which says that there exists one and only one parallel to a given line through a point not on that line. The negation of this statement, therefore, must mean one of two possibilities: either there exists *no* parallel to a given line through a point not on that line, or there exists *more than one* such parallel. If, Saccheri reasoned, he could show that either possibility would lead to a contradiction with the other nine axioms, then either must be false, and the truth of the Parallel Postulate would thereby be proved. It must be emphasized once again that we are not concerned here with the question of whether the Parallel Postulate is true in the physical world; we are only interested in its *logical* consistency with the other nine Euclidean axioms. Thus by "truth" we mean not physical truth, but logical truth.

Following this line of reasoning, Saccheri was able to show that the first alternative (no parallels to a given line) did indeed lead to a contradiction with the rest of Euclid's axioms, and he therefore eliminated it as a viable alternative. The second alternative, however, did not lead to any contradictions. It led, instead, to a series of strange results—strange, that is, to our common sense. For example, the sum of the angles of a triangle turned out to be *less than* 180°; what is more, the sum depended on the size of the triangle. (In ordinary geometry, of course, the sum is always equal to 180°, regardless of the size of the triangle.) Such strange results were enough to convince Saccheri that he had, indeed, found the contradiction which would prove the Fifth Postulate, and in 1733 he published his results in a book entitled *Euclid Vindicated from All Defects.* As it turned out, history would prove him both right and wrong: his methods were correct, but not his conclusions. Had he not been so eager to "vindicate" Euclid, he would have been credited with the creation of non-Euclidean geometry. Unfortunately, Saccheri lived in an era that was not yet ready for the breakthrough which, one hundred years later, would rock the foundations of mathematics. His work was soon forgotten.

The first to have the intellectual capacity to draw the right conclusions from these developments was Carl Friedrich Gauss (1777–1855). Popularly known as "the prince of mathematics,"

Gauss began to show his prodigious mathematical talents at a very young age. He mastered the art of calculation before he could read or write, and at the age of three he supposedly found an error in his father's bookkeeping. There is also the famous story about the ten-year-old Gauss who, when asked by his teacher to find the sum of the integers from 1 to 100, almost instantly came up with the correct answer: 5,050. To the teacher's astonishment, Gauss explained that he had noticed that by writing the sum once as $1 + 2 + 3 + \ldots + 99 + 100$, and again as $100 + 99 + \ldots + 3 + 2 + 1$, and then adding the two lines, each pair of numbers added up to 101. Since there were 100 such pairs, the sum of the two rows was 100×101 or 10,100, and the sum of each row was one half of this, or 5,050.

But the real turning point in Gauss's career came at the age of eighteen, when he proved that a regular polygon of 17 sides could be constructed with a straightedge and compass. Since the Greek era, the only regular polygons known to be constructible in this way were the equilateral triangle, the square, the pentagon, and the regular 15-gon, and the polygons derived from these by doubling the number of sides. Of these, only the triangle and the pentagon have a prime number of sides. Gauss showed that a regular polygon with a prime number of sides could be constructed with straightedge and compass if this prime number is of the form $N = 2^{2^n} + 1$, where n is a non-negative integer. For $n = 0, 1, 2, 3$, and 4, we get $N = 3, 5, 17, 257$, and 65,537, respectively—all primes.[3] Thus Gauss added three unknown constructible polygons to those known to the ancients: the regular polygons of 17, 257, and 65,537 sides. So impressed was Gauss with his feat that he requested a regular 17-gon to be engraved —

---

[3] The great French mathematician Pierre de Fermat (1601–1665) conjectured that the formula $N = 2^{2^n} + 1$ yields prime numbers for *every* non-negative integral value of n. As we have just seen, this is true for $n = 0, 1, 2, 3$, and 4. But in 1732, Leonhard Euler showed that for $n = 5$ we get the composite number $4{,}294{,}967{,}297 = 641 \times 6{,}700{,}417$, thereby disproving Fermat's conjecture. It is not known whether the formula yields any additional primes; thus it may be that there are other, as yet undiscovered, regular polygons constructible with a straightedge and compass. However, if such polygons exist, they must have a huge number of sides ($2^{26} + 1$, for example—which is composite— equals 18,446,744,073,709,551,617), making any *practical* construction totally out of the question.

The proof that primes of this form are the *only* ones for which the construction can be done was given in 1837 by Pierre Laurent Wantzel (1814–1848). Thus the condition that Gauss had found is both necessary and sufficient.

on his tombstone, in the tradition of Jacques Bernoulli's logarithmic spiral.[4] More significantly, however, was the fact that this achievement made Gauss decide on a mathematical career; up to that point he had seriously considered becoming a linguist. His subsequent contributions to mathematics were enormous, covering every field from number theory and algebra to celestial mechanics and the theory of functions of a complex variable (functions whose variables have the form $x + iy$, where x and y are real numbers and $i^2 = -1$). He gave the first full and correct proof of the Fundamental Theorem of Algebra, which says that every polynomial with complex coefficients has at least one complex root. In addition, he did a substantial amount of work in physics, particularly in electromagnetism, and the basic unit of the magnetic field is named after him.

It was at the age of fifteen that Gauss first turned his attention to the Parallel Postulate. Like all the others before him, he first tried to prove the postulate from the other Euclidean axioms, and, again like all his predecessors, he failed. He then began to suspect that the Parallel Postulate might be independent of the other axioms, and therefore not provable from them. But if this was so, then one could replace the Parallel Postulate, not by an equivalent postulate, as had been done so often before, but by a *contradictory* postulate, and still get a logically consistent geometry. Gauss realized, in other words, that the set of axioms one chooses as the foundation of a mathematical structure is, to a certain extent, arbitrary; change one or more of the axioms, and a different structure will emerge. Whether the new structure agrees with the "real" physical world is totally irrelevant; what matters is logical consistency alone. It was the inability to distinguish between these two issues that had hindered all previous attempts to prove the Fifth Postulate.

Now any scholar, having arrived at such far-reaching conclusions, would have rushed to publish them and establish credit for the discovery. Not Gauss! He had the rare trait of not publishing any of his work unless he was convinced beyond all doubt of its correctness. In this case, even though he correctly interpreted the failure to prove the Fifth Postulate, he never published his conclusions. It thus happened that two relatively unknown mathematicians claimed credit—independently—for the discovery of what is now known as non-Euclidean geometry. They were the

---

[4] Although his wish was never fulfilled, a regular 17-gon is engraved on the base of a monument in Gauss's honor in his native town of Brunswick, West Germany.

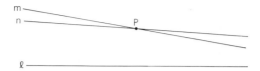

Figure 16.3. *The Parallel Postulate in hyperbolic geometry.*

Russian Nicolai Ivanovitch Lobachevsky (1793–1856) and the Hungarian Janos Bolyai (1802–1860).

The new geometry of Gauss, Lobachevsky, and Bolyai is based on the assumption that through a point P not on a straight line l, there are *at least two* straight lines m and n (in the plane of P and l) which do not meet l (Fig. 16.3). In other words, there exists at least two parallels to a given line through a point not on that line. One can now easily show that this assumption implies that there are, in fact, *infinitely many* parallels to l through P, namely, all the lines that could be drawn between m and n in Fig. 16.3. So this is the axiom that the three mathematicians decided to use instead of Euclid's Parallel Postulate; the other nine axioms remained unchanged. Their geometry, known as *hyperbolic geometry,* consists of all those theorems about two- and three-dimensional objects that can be derived from the new parallel postulate and the other nine Euclidean axioms. Many of these theorems are identical with the corresponding ones of Euclidean geometry— all those theorems whose proof does not depend on the Parallel Postulate. But others differ markedly from their Euclidean counterparts. For example, the sum of the angles of any triangle is always less than 180°, while in ordinary geometry it is equal to 180°. Even more surprisingly, the sum of the angles depends on the size of the triangle: the larger the size, the smaller the sum. It is not surprising that Saccheri had rejected such strange results; their very strangeness had convinced him that they must be false.

As we recall, Saccheri had examined the two possible alternatives to Euclid's Parallel Postulate—no parallels through a point to a given line, and more than one such parallel—and found the first to be inconsistent with the remaining nine axioms. But among these axioms, there is another, besides the fifth, which indirectly refers to infinity. This is the second axiom, which says that a straight line can be extended indefinitely in either direction. (In the original version: "To produce a finite straight line continuously in a straight line.") Once the possibility of replacing the Fifth Postulate by an alternative axiom had been accepted, it was not long before the second axiom, too, received a critical scrutiny. It was the Swiss mathematician Bernhard Riemann (1826–1866) who first replaced this axiom by one which says that a straight line is *unbounded.*

125

The distinction between "indefiniteness" and "unboundedness" is crucial; it is exemplified by the surface of the sphere, which has finite dimensions and yet is unbounded. Riemann's argument, like Gauss's before him, hinged on the inability of our senses to transcend the finite: we simply do not know what happens at infinity.

Having replaced the infinitude of the straight line by its unboundedness, Riemann could then replace the Parallel Postulate by the first of Saccheri's two alternatives—the one which Saccheri had found to be inconsistent with the other Euclidean axioms. In Riemann's geometry, there are *no* parallels to a given line through a point not on that line.[5]

The geometry based on Riemann's two assumptions and the remaining Euclidean axioms is known as *elliptic geometry.* As with hyperbolic geometry, some of its theorems are identical with their Euclidean counterparts, while others differ drastically from them. The sum of the angles of a triangle, for example, is always *greater* than 180°, and again depends on the size of the triangle. Moreover, any two similar triangles must also be congruent; that is, if they are identical in shape, they are also identical in size.

Of course, it is very difficult for our "common sense," being used to the rules of Euclidean geometry, to accept the validity of theorems such as these. Indeed, the first reaction to the new geometries was skeptical. Mathematicians were willing to acknowledge their logical consistency but continued to regard them as mere curiosities, devoid of any relevance to the real world. To facilitate their acceptance, several "models" have been devised. One of these, due to Felix Klein (1849–1925), is shown in Fig. 16.4. Imagine the entire universe to be confined to the interior of a circle. The concepts "point" and "line" are interpreted in the usual way, but we consider only points and line segments inside the circle—everything outside is "nonexistent." It is then clear that through a point P there are infinitely many "parallel" lines to a given line l, namely, all the lines lying outside the angle $\angle MPN$ in Fig. 16.5. (This is because none of these lines meets l inside the circle.) In Klein's model, therefore, the parallel postulate of hyperbolic geometry actually holds, while the other Euclidean axioms retain their validity. It is true that Euclid's second axiom, asserting the infinitude of the straight line, seems to be violated; but remember, we have confined the entire universe

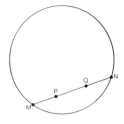

Figure 16.4. *The Klein model of non-Euclidean geometry.*

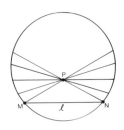

Figure 16.5. *The Klein model: there are infinitely many lines through* P *that are "parallel" to* MN.

[5] From this it becomes clear that in "vindicating" Euclid, Saccheri had tacitly assumed the infinitude of the straight line, though he may not have been aware of it.

to the interior of a circle, and it is therefore meaningless to extend a line indefinitely (in the Euclidean sense). We can, however, introduce the notion of *distance* into Klein's model, and define it in such a way as to make the distance infinite as we approach the boundary. Needless to say, such a "distance" would be quite different from the ordinary, Euclidean distance we use in daily life, but it can be made to fulfill all the properties associated with the concept of distance.[6]

Another model for hyperbolic geometry is due to Henri Poincaré (1854–1912). It is similar to the Klein model, except that the straight line segments are replaced by circular arcs that meet the boundary circle at right angles (Fig. 16.6). Once again, we can define the notion of distance in this model in such a way as to make the distance from any interior point to the boundary infinite—making the boundary forever inaccessible for the inhabitants of this universe. The Poincaré model recently became famous in quite an unexpected way: the Dutch artist Maurits C. Escher was so inspired by its unusual properties that he used it as a framework for one of his most beautiful works, *Circle Limit III* (Plate V).

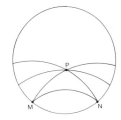

Figure 16.6. *The Poincaré model, later used by M.C. Escher in his* Circle Limit III.

But while such models may help in explaining the logical consistency of non-Euclidean geometry, they are to a certain extent artificial—they have very little to do with the real, physical world we live in. There is, however, one model of non-Euclidean geometry with which we are all familiar, and this is our own earth. In what follows we will assume that the earth is a perfect sphere, and that its inhabitants are primitive creatures who can move only along two directions—forward and backward, and right and left. For these creatures the world is two-dimensional; there is length and width, but no height.

What would our creatures mean by a "straight line"? Of course, words by themselves don't mean much until we give them a meaning through the properties of the things they describe. What do

---

[6] This distance is given by the formula $d_{PQ} = log\ (QM/QN)/(PM/PN)$, where *"log"* stands for "logarithm," M and N are the end points of the line segment through P and Q (Fig. 16.4), and QM, QN, PM, and PN denote the ordinary lengths of the respective line segments. Using the properties of the logarithm, one can show that as P approaches M (i.e., $PM \rightarrow 0$), d_{PQ} becomes infinite; the same is true when Q approaches N. Moreover, the formula obeys the "additive property" of distance: if P, Q, and R are three points on a line, then $d_{PQ} + d_{QR} = d_{PR}$. This follows from the well-known fact that the logarithmic function transforms a product into a sum. The ratio $(QM/QN)/(PM/PN)$ (acutally a ratio of ratios) is known as the *cross-ratio* of the points M, N, P, and Q and plays an important role in projective geometry.

127

we, three-dimensional beings, mean by saying that two points are connected by a straight line? We mean that the path connecting the two points has the shortest possible distance. We can demonstrate this by connecting the two points with a stretched rubber band: as long as the band lies entirely in a plane, it will follow a straight line. But suppose we are *not* on a plane; suppose we are confined to the curved surface of some three-dimensional solid—the surface of a globe, for instance. What is the shortest path between two points *on this surface* (no tunnels through the surface are allowed)? The answer is an arc of the *great circle* connecting the two points (Fig. 16.7). A great circle is a circle which cuts the sphere into two equal hemispheres; hence its center coincides with the center of the sphere. (Examples of great circles are the circles of longitude (meridians) and the equator.) Let us agree, therefore, to regard these great circles as the straight lines on the sphere. Indeed, for our two-dimensional creatures, great circles *are* the straight lines of their world.[7]

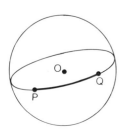

Figure 16.7. *A great circle: the shortest distance between two points on a sphere.*

Let us now imagine that our creatures decide to set out and explore their world, much in the same way as the explorers of the fifteenth and sixteenth centuries did, by embarking on sea voyages that will carry them to faraway, unknown continents. Naturally, they will follow a path which appears to them to be a straight line (actually an arc of a great circle), since this will bring them to their destination along the shortest route and, therefore, in the least time. Their voyage will at first carry them farther and farther away from home. But the time will come when the land they have reached will somehow look familiar: to their utter disbelief, they will have returned to their home port, albeit from the "wrong" direction. Unknowingly, they have circumnavigated their own globe—even though they had steadfastly followed a straight-line course!

On the surface of the sphere, then, Euclid's second axiom does not hold: straight lines are unbounded but finite. Does the Parallel Postulate hold? Is it possible to draw a parallel to a given line through a point outside that line? To answer this question, let us take as a straight line any great circle, say the equator. Then a parallel to this line can only mean that through a point not on the equator another great circle can be drawn which never meets the equator. But from the geometry of the sphere we know that two great circles *always* intersect; in fact, they intersect at two diametrically opposite points, or *antipodes,* on the sphere, such

---

[7] More generally, the shortest path on any given surface is known as the *geodesic* of that surface.

as the north and south poles (Fig. 16.8). We are therefore forced to conclude that on the surface of the sphere, parallel lines do not exist![8]

It is thus clear that on the surface of the sphere, neither Euclid's second axiom nor the Parallel Postulate hold. In fact, the conditions that do hold are precisely those which Riemann assumed for his elliptic geometry. But before we can declare that the sphere is a viable model for elliptic geometry, we must make one more observation. Euclid's *first* axiom states that one and only one straight line passes through two distinct points. But, as we have just seen, two great circles always intersect at a pair of antipodes. Therefore, through every pair of antipodes, more than one "straight line" passes—in fact, infinitely many, as the example of the meridians shows. Still, for all other points, the axiom does hold true, since through any pair of points other than antipodes, one and only one arc of a great circle passes. Now this is a perfect situation for a mathematician. Instead of marring the generality of our theory by having to make exceptions for some of its axioms, let's simply agree to regard any pair of antipodes as a *single point.* Strange as this may seem, it is not a new idea; we have used a similar strategy when we defined the points at infinity in projective geometry, or the single point at infinity in inversive geometry. As long as a new concept serves a good purpose, and as long as it does not contradict any previously defined concepts, we are free to accept it into our theory. With this agreement, we can now say that the sphere—and also the earth, to the extent that

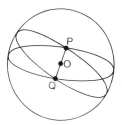

Figure 16.8. *Two antipodes (diametrically opposite points on a sphere) can be regarded as a single point in elliptic geometry.*

---

[8] It is one of the ironies of language that the term "parallel" is used in geography for the circles of latitude. True, the circles of latitude never intersect; but then, except for the equator, none of them are great circles, and hence they cannot be regarded as straight lines on the globe. The fact that a circle of latitude is *not* the shortest distance between two points is, for many people, a constant source of puzzlement. This is no doubt due to the fact that most of us are used to seeing the world as depicted on Mercator's map. As we have seen, this map shows the circles of latitude as horizontal, straight lines parallel to the equator, whereas great circles appear as curved paths. Thus the great circle route between Los Angeles and Tokyo, two cities with roughly the same latitude (34° and 36°, respectively), curves sharply northwards and reaches, at its northernmost point, a latitude of 48°, barely missing the Aleutian Islands! Flying one day from London to Chicago, I had the unforgettable experience of seeing below me the southernmost tip of Greenland, with its fjords and icebergs glittering in the August sun. Checking our route later on a globe, I was amazed how far north the great circle between the two cities reached. As it turned out, our pilot had taken the 747 even farther north than the great circle route, evidently taking advantage of the favorable trade winds at these high latitudes.

129

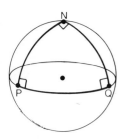

Figure 16.9. *The sum of the angles of a spherical triangle always exceeds 180°.*

it approximates a sphere—is a model for elliptic geometry. Indeed, some of the "strange" theorems in this geometry become quite obvious in our model—for example, the fact that the sum of the angles of a triangle always exceeds 180° (Fig. 16.9).

Similar models exist for the hyperbolic geometry of Gauss, Lobachevsky, and Bolyai, in which there are *infinitely many* parallels to a given "straight line." (Again, by a "straight line" we mean the shortest route between two points on the given surface; see footnote 7.) One of these models, the so-called pseudosphere, is shown in Fig. 16.10. But why do we call them "models" rather than actual spaces in which the laws of non-Euclidean geometry hold? The reason is that we are three-dimensional beings; we can move in any direction—including up and down. We are able to view our two-dimensional creatures from the vantage point of the third dimension, and thus explain their two-dimensional spherical geometry in terms of our ordinary, *three-dimensional* Euclidean geometry. In fact, there is a branch of mathematics, spherical trigonometry, which deals with the properties of triangles and other shapes on the surface of the sphere. (This topic, which is of a rather technical nature, is studied in courses on surveying and navigation.) With the help of spherical trigonometry, we can derive all the properties of the surface of the sphere from those of ordinary solid geometry. In other words, the axioms of two-dimensional elliptic geometry become *theorems* in three-dimensional Euclidean geometry. But whatever three-dimensional privileges we humans are endowed with, these privileges are denied our creatures. To their knowledge and experience, there *are* only two dimensions. If a stranger from the third dimension told them that their straight lines are really circles, they would be puzzled indeed; for a circle must have a center, and the common center of all great circles is at the earth's center, which is forever inaccessi-

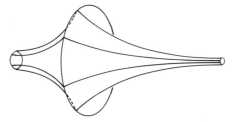

Figure 16.10. *The* tractroid. *This surface is a model for hyperbolic geometry, since on it there are infinitely many parallel lines to any given line. (A "line" here means the shortest distance between two points on the surface.) Also known as the* pseudosphere, *this surface shares with the sphere the property that it everywhere has the same curvature: for a sphere, this constant curvature is positive, while for the tractroid it is negative.*

ble to them. So from their point of view, they live in a two-dimensional world in which the laws of non-Euclidean geometry prevail. This is precisely what we mean by saying that *the surface of the sphere is a Euclidean model for the non-Euclidean elliptic plane.*

But mathematics is seldom totally isolated from our physical environment; directly or indirectly, it derives its ideas from the real universe of which we are a part.[9] Soon after the logical soundness of non-Euclidean geometry had been acknowledged, the question arose as to whether it may have some relevance to the material world around us. More specifically: Is it possible that physical space, on a large scale, obeys the laws of non-Euclidean geometry?

The odds against such a possibility seem to be overwhelming indeed. Everything in our daily experience suggests that physical space is Euclidean; after all, our technological achievements are all based on Euclidean geometry, and this fact alone should have decided the matter once and for all. But then, our experiences are necessarily confined to the immediate world around us; we have no direct way of knowing what happens at the far reaches of the universe. Gauss himself expressed these thoughts in a letter to his friend Schumacher: "Finite man cannot claim to be able to regard the infinite as something to be grasped by means of ordinary methods of observation." Gauss, who was a physicist as much as a mathematician, was one of the very few modern scientists who felt equally at home in theory and in practice, and it did not escape his mind that his ideas about non-Euclidean geometry could have profound implications for physics. He therefore decided to settle the matter by an experiment.

To be sure, it would be quite hopeless to determine directly which of the three parallel postulates is the true one (from a

---

[9] There are, to be sure, two opposing schools of thought on this matter. One school consists of the "pure" mathematicians, those who pursue their profession for the sheer intellectual satisfaction it offers; the other is the "applied" school, whose members are mainly concerned with the applications of mathematics to science and technology. The first school maintains that mathematics need not have any relevance to the physical world. (Some go even further, claiming that the more removed mathematics is from practical applications, the better for the profession!) According to this school, a mathematician is free to create his or her own "real" world, provided it is logically consistent (i.e., free from internal contradictions). Non-Euclidean geometry, with its intellectual, almost aloof attitude, was the epitome of pure mathematics. Lobachevsky himself, in his epoch-making paper of 1829, called his new geometry "imaginary," implying that it was to be regarded as a purely logical, axiomatic discipline and not as a description of physical space. By contrast, the invention of calculus, which was in direct response to the new discoveries in physics and astronomy, was the great achievement of applied mathematics.

physical point of view), because we simply cannot go out to infinity and check whether or not parallel lines meet. We can, however, use some of the theorems which result from accepting one or the other of the various parallel axioms, and most of these theorems deal with figures of finite size. For instance, we have seen that the sum of the angles of a triangle is always less than 180° in Guass's hyperbolic geometry, and always greater than 180° in Riemann's elliptic geometry. (It is, of course, *equal* to 180° in Euclidean geometry.) So by measuring the angles of a very large triangle, it might just be possible to decide which of the three cases actually holds. Gauss attempted to do just that. He stationed surveyors on three mountain summits at considerable distance from each other, and had them measure the angles between their lines of sight. The results were disappointing: the sum of the angles turned out to be almost exactly 180°, any deviation from this value falling entirely within the limits of accuracy of the measuring instruments. Hence the experiment settled nothing. Gauss soon realized the futility of any attempt to settle the issue by a terrestrial experiment; only distances on a celestial scale might carry some hope of success, and the accuracy and range of the astronomical instruments of the time were far from sufficient for such a task.[10]

The final verdict on the issue had to wait one more century. In 1916, a young and relatively unknown German physicist published a new theory of gravitation, in which space and time were unified into one four-dimensional entity. In his theory, straight lines were defined as the paths of rays of light; since no material object can travel faster than the speed of light (approximately 300,000 km/sec in vacuum), a light ray represented the shortest

---

[10] It should be clearly understood that the issue was not to confirm the non-Euclidean nature of the earth's surface when regarded as a *model* for elliptic geometry—there had been no argument on this point—but rather to test the nature of *physical, three-dimensional* space.

It must also be pointed out that the three geometries—plane (Euclidean), elliptic, and hyperbolic—are nearly in agreement when the figures involved are very small. For example, the sum of the angles of a spherical triangle, though always greater than 180°, gets closer to 180° as the size of the triangle shrinks to zero. Since in daily life we are seldom aware of the sphericity of the earth, it is not surprising that Euclidean geometry has been held for so many years as the only true geometry of our physical world. It is a testimony to the intellectual insight of the Greek classicists that they had erected the science of mathematics on the foundations of Euclidean geometry. Even if some of them had suspected that the earth might be round, from a practical viewpoint their world was flat, and they created the right type of geometry to describe this world.

route between two points ("events") in this space–time universe. Moreover, the theory predicted that light rays should be bent in the presence of a strong gravitational field, such as that created by a massive star. But this meant that space–time must be curved— much in the same way a taut membrane is curved when a heavy object is placed on it—and that the curvature at every point depends on the strength of the gravitational field at that point. To describe such a four-dimensional curved space mathematically, the young physicist was looking for some kind of non-Euclidean geometry, and he found it in Riemann's geometry, which allows for space to have a variable curvature.

The new theory, with its strange predictions, aroused considerable controversy in the scientific community, but in 1919 an opportunity presented itself to put the theory to a test. On May 29 of that year, a total solar eclipse was scheduled to occur, and expeditions were sent to Brazil and to the west coast of Africa to photograph it. When the stars on the photographic plates were compared to the same area of the sky photographed half a year later (i.e., during nighttime), the positions of the stars whose lines of sight passed very close to the sun were found to have shifted from their normal positions by just the amount predicted by the theory. News of the successful test flashed throughout the world, and overnight the creator of the theory became world famous. His name was Albert Einstein, and his theory, the general theory of relativity.[11]

Since that day, general relativity—which is really a geometric theory of gravitation—had been put to the most demanding tests (most recently using signals from spacecraft), and it has withstood them all. General relativity, therefore, can be said to be the final triumph of a mathematical idea which, in its infancy, was no more than an intellectual exercise. This tremendous triumph, however, must not obscure the fact that non-Euclidean geometry remains, first and foremost, a mathematical theory. In retrospect, its most significant achievement was not the central role it would play in modern physics, but the intellectual breakthrough it would bring about. This breakthrough can perhaps best be compared to the Copernican revolution some three centuries earlier. Just as Copernicus had removed the earth from its lofty position as the center of the universe, so did Gauss, Riemann, and their followers remove Euclidean space from the supreme role it had played in geometry.

---

[11] The dramatic developments which followed this historic event have been described many times. See, for example, *Einstein: The Life and Times* by Ronald W. Clark, Avon Books, New York, 1971.

The analogy goes even further. Copernicus had found it inconceivable that the entire universe should revolve around a tiny planet called Earth; Gauss questioned our right to pass judgment about the behavior of geometric space far away from us. Both revolutions thus owe their origin to a common root—the infinite. Like every new idea, both were at first met with skepticism, resistance, and in Copernicus's case, outright hostility. But once accepted, they changed the course of history. They have shaken our belief in the power of science to show us the way to the ultimate, absolute truth, because no such truth exists. Absolute truth had to be replaced by relative truth, one that depends on the premises we set at the start. Very gradually and imperceptibly at first, the journey had begun which, a century later, would culminate in our modern view of the universe.

Die zwei Parallelen

by

Christian Morgenstern (1871–1914)

Es gingen zwei Parallelen
ins Endlose hinaus,
zwei kerzengerade Seelen
und aus solidem Haus.

Sie wollten sich nicht schneiden
bis an ihr seliges Grab:
Das war nun einmal der beiden
geheimer Stolz und Stab.

Doch als sie zehn Lichtjahre
gewandert neben sich hin,
da ward's dem einsamen Paare
nicht irdisch mehr zu Sinn.

War'n sie noch Parallelen?
Sie wusstens selber nicht,
sie flossen nur wie zwei Seelen
zusammen durch ewiges Licht.

Das ewige Licht durchdrang sie,
da wurden sie eins in ihm;
die Ewigkeit verschlang sie,
als wie zwei Seraphim.

Aesthetic Infinity   Part III

17 Rejoice the Infinite!

Oh moment, one and infinite!

— Robert Browning (1812–
1889), *By the Fireside*

From times immemorial, man has aspired to the infinite. The first known attempt to reach infinity occurred in Babel and is told in Genesis: "And they said: 'Come, let us build a city, and a tower, with its top in heaven.'" Their attempt was doomed to fail, for God, fearing that their aspirations may be too high, "confounded their language, that they may not understand one another's speech" (Genesis 11:4). Ever since, the Tower of Babel has become an allegory of the infinite—or of man's futile efforts to reach it.

Having failed to reach the infinite physically, man has turned inward to reach it spiritually. Poets, artists, and philosophers of all times have been obsessed with it; some feared it, others rejoiced in it, but few could escape its intoxicating grip. "The eternal silence of these infinite spaces terrifies me" lamented Blaise Pascal (1623–1662), mathematician, physicist, and philosopher, in his characteristically gloomy outlook of the world. The Jewish philosopher Martin Buber (1878–1965) was so terrified by the infinite that he contemplated committing suicide to avoid it:

> *A necessity I could not understand swept over me: I had to try again and again to imagine the edge of space, or its edgelessness, time with a beginning and an end or time without beginning or end, and both were equally impossible, equally hopeless . . . Under an irresistible compulsion I reeled from one to the other, at times so closely threatened with the danger of madness that I seriously thought of avoiding it by suicide.* [1]

---

[1] *Martin Buber's Life and Work: The Early Years 1878–1923,* by Maurice Friedmann, E.P. Dutton. New York, 1981.

Contrast this with William Blake's (1757–1827) optimistic lines in *Auguries of Innocence:*

> To see a world in a grain of sand
> And a heaven in a wild flower,
> Hold infinity in the palm of your hand
> And eternity in an hour.

And Friedrich von Schiller (1759–1805) ended his hymn for the creation with this tribute to the infinite:

> Thou sail'st in vain—Return! Before thy path, INFINITY!
> And thou in vain!—Behind me spreads INFINITY to thee!
> Fold thy wings drooping,
> O Thought, eagle-swooping!—
> O Fantasy, anchor!—The Voyage is o'er:
> Creation, wild sailor, flows on to no shore![2]

Vincent van Gogh (1853–1890) spoke of the "vertigo of the infinite." To quote his biographer, Henri Scrépel: "From the rock of Montmajourn [in southern France] he discovered, not the misty plain of the North, with its blurred outline, but an immense and shadowless expanse in which the most impalpable details could be discerned. *'I am painting the infinite,'* he had written one day. He could see it unfolding before him: the infinitely great in the plain stretching as far as the eye could see; the infinitely small in the proliferation of fields, olive-trees, vines and stones, the myriads of microscopic pores in the face of the earth."[3]

Explorers, too, have experienced this intoxication with the infinite—the infinite of vast spaces, of endless lands on which no human has set foot before, of limitless oceans where the elements are given free rein. Sir Francis Chichester (1902–1972), aviator and sailor, tells of his exhilaration while flying his tiny plane en route to Australia: "Above, infinite space, illimitable emptiness, with only the sun shining brazenly, eternally . . . What complete isolation, solitariness!"[4] Alan Moorehead described the vast, unexplored deserts of central Australia in *Cooper's Creek.* Recounting the hardships of the Burke expedition, the first to cross the continent, he says: "They were aliens in this hard, indifferent country, the gaol of interminable space"; whenever they turned, they could only see "empty land stretching silently into infinity." To some,

---

[2] From *The Greatness of Creation,* translated from the German by Sir Edward Bulwer Lytton, Thomas Y. Crowell & Co., New York.
[3] *Van Gogh,* by Henri Scrépel, Paris, 1972.
[4] *Solo Flight to Sydney,* by Francis Chichester, Stein and Day, New York, 1982.

the infinite had a healing power, lifting their souls from the pettiness and worries of daily life. "I have been accustomed to the desert, to infinite spaces, where I have not had to worry about the trivialities that stifle man."—said the German explorer of the Sahara, Heinrich Bart (1821–1865).

The use of silence as a metaphor of the infinite is a recurring theme among authors. Wassily Kandinsky (1866–1944), the Russian-born French artist, spoke of "A great silence, like a cold indestructible wall going on into the infinite." The British physicist John Tyndall (1820–1893) compared the mind of man to a "musical instrument with a certain range of notes, beyond which in both directions we have an infinitude of silence." And who can remain unmoved by the final notes of Gustav Mahler's (1860–1911) *Das Lied von der Erde,* when the singer whispers the words *Ewig . . . ewig . . .* (eternity . . . eternity . . .), and the music dies away to nothingness, as if anticipating the composer's own death.

If silence is the audible metaphor of the infinite, blue is its sensual symbol. John Ruskin (1819–1900), the English art critic, spoke of the "blue of distance." Blue was the favorite color of the Spanish artist Joan Miró (1893–1983); in *Towards the Infinite* he used it to portray a limitless void, through which a thin line streaks upwards and soars to infinity. Undoubtedly it was the blue of the sky that brought about this association with infinite spaces. "Vast expanses of sky and ocean accomodate the spirit craving release through a sense of infinity," wrote the author of a recent book on color when discussing the "infinity of blue."[5]

From these general remarks let us now turn to some specific instances where the idea of infinity has inspired artists to create works of great aesthetic appeal.

---

[5] *Color,* edited by Donald Pavey, Marshall Editions Limited, London, 1980.

138

The Möbius Strip 18

I am of the opinion that it is possible to develop an art largely on the basis of mathematical thinking.

— Max Bill (b. 1908)

At one time a famous mathematical curiosity, later a source of inspiration for artists, the strange properties of the Möbius strip have fascinated professionals and laymen alike ever since its discovery in 1865. Named after its creator, the German mathematician and astronomer August Ferdinand Möbius (1790–1868), it was the embryo of an entirely new branch of mathematics known as topology, the study of those properties of a surface which remain invariant when the surface undergoes a continuous deformation.

Take a rectangular strip of paper, as in Fig. 18.1a, and bend it to form a ring-like object. If you bend it in the usual way, by connecting points A with B and C with D, you get an ordinary ring—a circular, endless strip that has an inside and an outside (Fig. 18.1b). But if you first give it a half-twist, and then connect points A with D and C with B, a Möbius strip will result (Fig. 18.1c).

The first thing we notice about the Möbius strip is that it has only *one side:* we can go from any point on one side of the strip to any point on the "other" side along a continuous path, without ever penetrating the surface or going over its edge! This, of course, is impossible with an ordinary, two-sided strip. Thus a Möbius strip has neither an "inside" nor an "outside," just a single, continuous surface. Furthermore, while an ordinary endless strip has two edges, a Möbius strip has but one: starting at any point on the upper edge and moving along this edge, we reach, after one circuit, a point opposite the starting point on the lower edge; after another circuit, we are back at the starting point. We must

Figure 18.1. *Construction of a Möbius strip.*

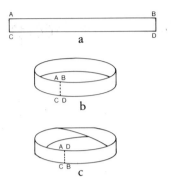

complete *two* circuits to go around a Möbius strip—again in contrast to an ordinary endless strip.

Still more surprises await us when we try to cut a Möbius strip into two halves. If we cut an ordinary endless strip along its centerline, we get two separate identical strips, each one half as wide as the original strip. But if we do the same with a Möbius strip, we get a single, continuous object—a ring that has *two* half-twists built into it. This object is no longer a genuine Möbius strip, since it has two distinct edges and two distinct sides—an inside and an outside. But if we cut a Möbius strip lengthwise along a line one-third the way across, we get two intertwined rings, one a true Möbius strip, the other the two-sided ring with two half-twists! Little wonder that the Möbius strip has become one of the most intriguing mathematical creations ever conceived, a kind of geometric magic work whose secrets are well hidden behind its twisting surface.

It was only natural, therefore, that the Möbius strip would draw the attention of the world of art. Among the artists attracted by its aesthetic possibilities was the Swiss sculptor Max Bill (b. 1908), who discovered it in 1935: "I created a single-sided object by searching for a solution of a hanging sculpture, turning in the rising air. My search was neither scientific nor mathematical, but purely aesthetic . . . I named my sculpture *Endless Ribbon."* Evidenctly Bill was unaware that his creation had been known to mathematicians for nearly a century, and his disappointment upon learning of the fact was great. To quote Bill: "Sometime later I was informed that my creation, which I thought I had discovered or invented, was only an artistic interpretation of the so-called Möbius strip, and theoretically identical to it . . . I was shocked by the fact that I was not the first one to discover this object. I therefore stopped all further research in this direction for a while." Some years later, however, he turned again to topological problems and single-sided surfaces. By his own testimony, it was only

140

in 1979, almost 120 years after Möbius described his "unilateral polyhedra," that Bill was confronted with the original explanation of the strip. But, he added, "what I could not find in Möbius' explanation is of primary importance to me: the philosophical aspects of these surfaces as symbols of infinity." So Bill set himself to the task of realizing their aesthetic potentialities. The result were several works of exquisite beauty, one of which we show here (Fig. 18.2).

While Max Bill's art, with its stark, clean lines, is pure mathematics translated into art, another contemporary artist has used the Möbius strip in an "applied" sense. Maurits C. Escher (1898–1972) became aware of the Möbius band through a casual encounter with an English mathematician, whose identity, unfortunately, he could not remember later. The encounter, evidently, was a

Figure 18.2. *Max Bill: Endless Ribbon from a Ring II (1947/48), with the artist's permission.*

141

Figure 18.3. *M.C. Escher:* Möbius Strip I *(1961).* © *M.C. Escher Heirs c/o Cordon Art—Baarn— Holland.*

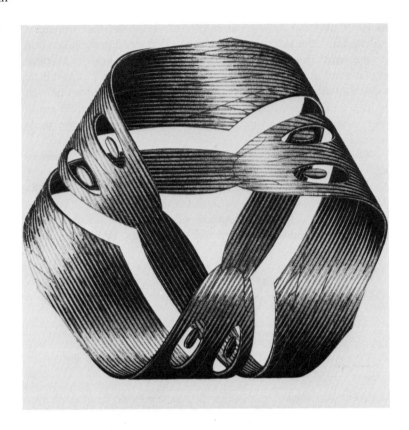

fruitful one, for it inspired Escher to create three works based on the Möbius motif. With his characteristic bent for the grotesque, Escher's creations teem with life: giant ants climbing their way up one side of a ladder shaped as a Möbius strip, only to find themselves descending the "other" side in an endless cycle (Plate I); a pair of abstract creatures—snakes, perhaps—chasing each other along the seemingly separated parts of a Möbius band cut along its length (Fig. 18.3); and a parade of red and grey horsemen, one group the exact mirror image of the other, marching in opposite directions along the two sides of an endless, twisted strip (Plate II).[1] Escher, a genius in portraying the ambiguities

---

[1] This is not a true Möbius strip, since it has two sides and two edges; in fact, it is the object obtained when a Möbius strip is cut lengthwise along its centerline. It can also be obtained by giving the rectangular strip of Fig. 18.1a two half-twists before joining the end points. To add to the complexity, Escher's strip in *Horsemen* is connected at the center of the picture, bridging the two separate sides and enabling the two groups of horses to meet.

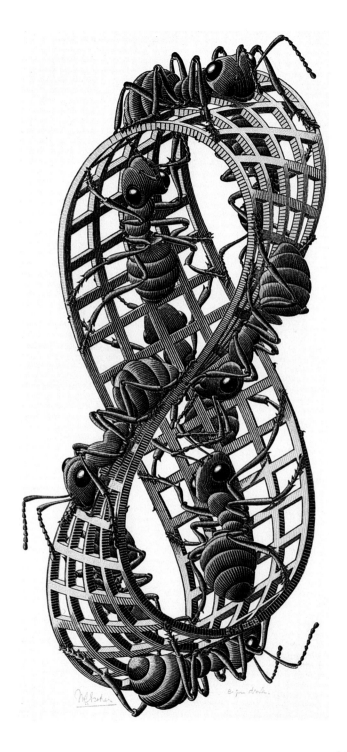

Plate I. *M.C. Escher: Möbius Strip II* (1963). © *M.C. Escher Heirs c/o Cordon Art—Baarn—Holland. Collection Haags Gemeentemuseum—The Hague.*

143

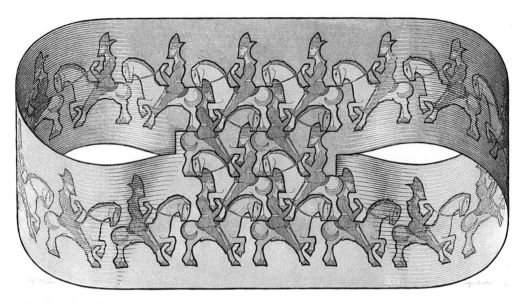

Plate II. *M.C. Escher: Horsemen* (1946). ©
M.C. Escher Heirs c/o Cordon Art—Baarn—Holland. Collection Haags Gemeentemuseum—The Hague.

Plate III. *M.C. Escher: Regular Division of the Plane with Animals.* © *M.C. Escher Heirs c/o Cordon Art—Baarn—Holland. Collection Haags Gemeentemuseum—The Hague.*

144

Plate IV. *M.C. Escher:*
Smaller and Smaller
(1956). © *M.C. Escher*
Heirs c/o Cordon Art—
Baarn—Holland.
Collection Haags
Gemeentemuseum—The
Hague.

145

Plate V. *M.C. Escher:*
Circle Limit III (1959).
© *M.C. Escher Heirs c/o*
Cordon Art—Baarn—
Holland. Collection Haags
Gemeentemuseum—The
Hague.

Plate VI. *M.C. Escher:*
Whirlpools (1957). ©
M.C. Escher Heirs c/o
Cordon Art—Baarn—
Holland. Collection Haags
Gemeentemuseum—The
Hague.

147

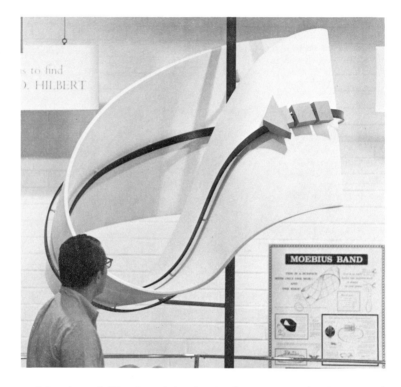

Figure 18.4. *A model train winding its way along a huge Möbius band. This fascinating show is part of the IBM–sponsored exhibition* Mathematica: The World of Numbers and Beyond *by Charles and Ray Eames, which has made a tour of several major science museums in the U.S.A. Photograph by Charles Eames, with permission.*

and ironies of life, found in the Möbius strip a fertile ground for his creative talents.

In the science fiction story, "A Subway Named Möbius,"[2] the plot centers around the mysterious disappearance of train No. 86 from the Boston subway system. The system had just been inaugurated the day before, but now No. 86 has vanished without a trace. Indeed, numerous people had reported hearing the train rush by right above or below them, but none had actually seen it. When all attempts to locate the train have failed, Roger Tupelo, the Harvard mathematician, calls up Central Traffic to offer a startling theory: the subway system is so complex that it may have become part of a single-sided surface—a Möbius band—and the missing train may, at this very moment, be running its normal course on the "other" side of the band. To the utter consternation of the city officials he patiently explains the topological singularities of such a system. And sure enough, after a while—ten weeks, to be exact—the missing train reappears, its passengers all well, if a bit tired.

---

[2] By A.J. Deutch (1950), in *Where Do We Go From Here?*, edited by Isaac Asimov, Fawcett Crest, Greenwich, Connecticut, 1972.

148

The Magic World of Mirrors 19

See now the eternal Virtue's breadth and height,
Since it hath made itself so vast a store
Of mirrors upon which to break its light,
Remaining in itself one, as before.

— Dante Alighieri (1265–1321), *Paradiso,* Canto 29

Everyone has, at one time or another, been fascinated by mirrors. Even animals. I remember once watching a cat staring with puzzlement at a mirror, no doubt wondering who that other fellow cat behind the shiny surface was. The laws of optics play here a subtle trick: a ray of light falling on a mirror is reflected at exactly the same angle at which it hit the mirror,[1] creating the illusion that a hidden object appears from behind the mirror (Fig. 19.1). From the dawn of history, mirrors have been regarded as treasured possessions; we find them in the tombs of Egyptian royalties— evidently put there to preserve the beauty of the deceased for

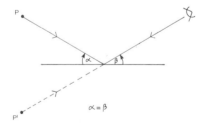

Figure 19.1. *The law of reflection.*

---

[1] In more formal language: "The angle of incidence equals the angle of reflection." This is the law of reflection, one of the two fundamental laws of geometric optics. (The other is the law of refraction, governing the passage of light from one medium to another.) The law also says that the incident ray, the reflected ray, and the normal to the plane of the mirror all lie in the same plane.

Figure 19.2. *Two inclined mirrors.*

future generations—as well as in the vast halls of the palaces of Baroque Europe, where they entertained the nobility with their glittering reflections. In our own time they are being used by department stores and restaurants as a commercial gimmick, creating the illusion of a large space where the actual room is rather small. And in the form of curved mirrors, they provide endless fascination for the youngsters in a magic house.[2]

Take two mirrors and place them at an angle, as in Fig. 19.2. An object placed anywhere between them is reflected in each of the mirrors (or in their extensions), producing two images. Each of these images can now be regarded as a new object, which is again reflected in each mirror, and so on. It seems as though the repeated reflections would multiply endlessly, creating an infinite number of images. But it may happen that some of the secondary reflections will coincide, in which case the process will come to an end. Figure 19.3 is a cross section of two mirrors forming an angle of 60°. The solid lines A and B represent the actual mirrors, the thin lines are their extensions, and the broken line C is the reflection of either mirror in the other. Now place an object K anywhere between A and B (in the figure, the shaded object is touching B, but this is done only to make it easier to follow its reflections). We denote by A_1 and B_1 its reflections in the mirrors A and B, respectively. A_1 is now reflected in B, producing a new image A_2; likewise, B_1 gives rise to a new image B_2 by its reflection in A. Finally, each of these secondary images is again reflected in the original mirrors, giving rise to A_3 and B_3. But from the geometry of reflections we can easily convince ourselves that these last two images coincide, so that no further images are produced. From the original object we thus obtain five new images; together with the original they are arranged in three pairs, forming the vertices of an equilateral triangle.

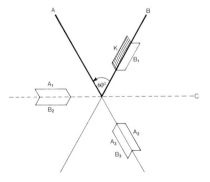

Figure 19.3. *Repeated reflections in two mirrors forming a 60° angle, resulting in a symmetric pattern.*

150

Thus, for the particular angle of 60°, we get a pattern character-ized by a high degree of symmetry: the pattern can be reflected in any of the lines *A, B,* and *C,* or it can be rotated through any of the angles 120°, 240°, and 360°, without changing the pattern as a whole. We say that the pattern has a *threefold rotation symmetry* and a *three-axis reflection symmetry.* This result can be gener-alized: if the mirrors form an angle α, the resulting pattern will have an n-fold rotation symmetry and an n-axis reflection symme-try, where $n = 360°/2\alpha$.[3]

It is this principle—the creation of a symmetric pattern from an object placed between two inclined mirrors—that is the basis for the *kaleidoscope.* Invented in 1816 by the Scottish physicist Sir David Brewster (1781–1868), the kaleidoscope became the wonder toy of young and old alike. (The name is derived from the Greek words *kalos* = beautiful, *eidos* = form, and *scope* = to view.) It consists of a tube inside which two mirrors are held in place, usually at an angle of 60°. At one end of the tube little chips of colored glass are placed, and these, with their multiple reflections in the mirrors, can be viewed through a hole at the other end. So simple was the construction of the instrument that thousands were copied without permission, depriving its inventor of the royalties due to him. Apparently the instrument has lost nothing of its appeal even today; in its modern version the mirrors are replaced by a small multilens, and the object is any scene at which you choose to direct the tube.[4]

In an actual kaleidoscope the angle α is fixed, but we can also experiment with two free mirrors and hold them to each other

---

[2] More about the fascination of mirrors can be found in *The Ambidextrous Universe* by Martin Gardner, Charles Scribner's Sons, New York, 1979.

[3] This formula gives the number of symmetry elements only when n has an integral value. For this to happen, the angle α must be restricted to certain values (180°, 90°, 60°, 45°, 36°, etc., corresponding to $n = 1, 2, 3, 4, 5, \ldots$). If α has any other value, n will be a fraction, or even an irrational number, resulting in the loss of some or all of the rotational symmetry. Also, if the object itself already possesses reflection symmetry, and if we place it symmetrically with respect to the mirrors, then additional symmetries result. For example, for the case $\alpha = 60°$, if the object is the letter I, and if we align it along the bisector of the two mirrors, the resulting pattern will have the symmetry of a regular hexagon, i.e., a sixfold rotation symmetry and a six-axis reflection symme-try. For a fuller discussion of these possibilities, see H.S.M. Coxeter, *Introduction to Geometry,* listed in the bibliography.

[4] Instructions for constructing a simple kaleidoscope can be found in *Pho-tography as a Tool,* Life Library of Photography, Time-Life, New York, 1970, p. 208.

151

at different angles. It is fascinating to watch the number of images multiply as we gradually decrease the angle; more and more images emerge seemingly from nowhere, until their number is so large that they become blurred. (This is partially due to imperfections in the mirrors, creating internal reflections which become more pronounced as the line of sight increasingly approaches the plane of the mirror.) As $\alpha \rightarrow 0$, the number of images becomes infinite; but then, of course, there is no more room left to place an object between the mirrors, and no more room for us to watch its reflections. The secrets of infinity are well guarded by our mirrors!

There is, however, a way to get around this situation, and that is to place oneself between two *parallel* mirrors, separated by a constant distance d. This is the familiar sight we encounter at the barbershop, where you can see yourself reflected back and forth in an endless display of mirrors, showing alternately your face and back. Well, *almost* endless, for once again we are prevented from actually seeing infinity. For one, your own head will always be in your line of sight, blocking the view straight ahead; and secondly, the slightest deviation from exact parallelism will turn the two mirrors into a giant kaleidoscope, causing the images to curve along a vast circle until they get out of view.

Imagine now that you are in a room all four of whose walls are mirrors. Figure 19.4 shows a cross section of such a room. As before, the solid lines denote the actual mirrors, forming a

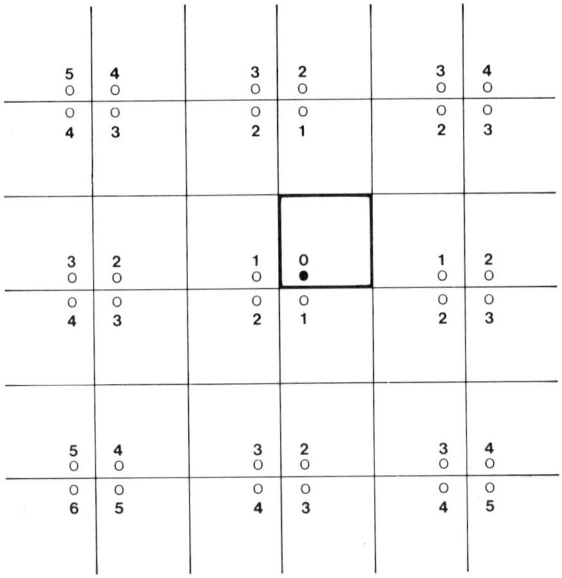

Figure 19.4. A two-dimensional mirrored room. The small numbers denote the sequence of repeated reflections.

square, while the thin lines are their repeated reflections. Placing an object anywhere inside this square produces a kaleidoscopic effort around the corner nearest to the object; but in addition, a new element enters the picture. Since the repeated reflections of each pair of opposite mirrors create an infinite lattice of squares (each of which can be regarded as a new mirrored square), the original kaleidoscopic pattern is repeated again and again, spreading into the entire plane. The result is a tessellation of the plane by squares, each square being "ornamented" around one of its corners by the images of the original object. A similar effect can be achieved with three mirrors forming an equilateral triangle, resulting in a triangular tessellation.

From here it takes just one step to imagine an entire room of mirrors—floor, walls, and ceiling. Such a room, complete with a table and chairs all made of mirrors, is exhibited at the Albright-Knox Gallery in Buffalo, New York. It is actually a work of art, created in 1966 by the Greek-born American artist Lucas Samaras (b. 1936). Once inside, you get as close to seeing infinity as you can ever hope for—and a dizzying experience it is: you feel as though you are standing at the brink of a bottomless pit, ready to swallow you if you make a careless move.

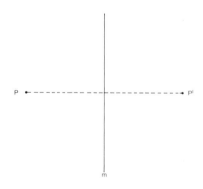

Figure 19.5. *Geometric definition of reflection: the line* m *is the perpendicular bisector of* PP'.

We can now replace the actual mirror by a geometric abstraction—a line *m* which we will call the *axis of reflection,* or the *mirror axis* (the term *symmetry axis* is also used). By "reflection" in this line we will mean a transformation that carries every point *P* in the plane to a point *P'* such that the axis of reflection is the perpendicular bisector of the segment *PP'* (Fig. 19.5). In this way we can regard reflection as a purely geometric concept, independent of any optical considerations. This sort of "mathematical optics"

153

has become a fruitful field of research, abounding with many beautiful results.[5] In what follows we will adopt this viewpoint.

---

[5] An interesting problem, as yet unsolved, is the following: Given a convex polygon whose inner sides are one-dimensional mirrors, we can illuminate the entire interior from a source of light placed anywhere inside the polygon (even without the mirrors). Can we do the same for *any* mirrored polygon—convex or not? The answer seems to be "yes," since the reflected rays of light, bouncing back and forth from the walls like billiard balls on a pool table, seem to be able to reach every interior point of the polygon. But this has not been proved so far. Interestingly enough, if the polygon is replaced by an arbitrary closed curve, it is known that some shapes cannot be entirely illuminated from an interior source of light, no matter where it is placed. This is rather surprising, since any curve can be approximated as closely as we please by straight-line segments, i.e., by a polygon. For details, see *Tomorrow's Math: Unsolved Problems for the Amateur* by C. Stanley Ogilvy, Oxford University Press, New York, 1972, p. 59.

Horror Vacui, Amor Infiniti 20

If in the infinite you want to stride,
Just walk in the finite to every side.

— Johann Wolfgang von Goethe
(1749–1832)

We now arrive at one of the most beautiful applications of mathematics to art—a study of the various possibilities of filling the plane with infinite repetitions of a single artistic design. Such infinite patterns have captured the imagination of artists and craftsmen since the earliest recorded time and have provided the framework for the exquisite abstract art of the Moslems. It is in these designs that the interplay between geometry and art reaches its supreme level.

As we have seen, the repeated reflections of an object in two or more mirrors creates a symmetric pattern, even if the object itself possesses no symmetry. If we examine this process closely, we find that three types of movements, or transformations, take place in the plane of the figure: reflections, rotations, and translations. As defined before, a *reflection* in a line m is a transformation that carries every point P to a new point P' such that m is the perpendicular bisector of the segment PP'. This definition is purely geometric—it does not depend on any optical reflection actually taking place. The other two transformations can also be defined in purely geometric terms, independent of any "physical" considerations. A *rotation* (or *turn*) in the plane about a fixed point O carries every point P to a new point P' such that $OP = OP'$; the angle of rotation is the same for all points (Fig. 20.1). A *translation* carries every point P a fixed distance d in a given direction; therefore, all points of the plane are displaced, or translated, by the same amount and in the same directions (Fig. 20.2).

These three transformations are the basic tools needed to study the symmetry elements of any figure. A *symmetry element,* or *symmetry operation,* is any transformation that leaves a figure invariant

155

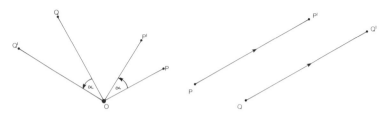

Figure 20.1. *Rotation.*   Figure 20.2. *Translation.*

(unchanged) as a whole, even though its individual parts may permute. For example, an equilateral triangle has six symmetry elements: three reflections in its three bisectors and three rotations, each of 120°, about the point of intersection of these bisectors (Fig. 20.3). Each one of these operations leaves the triangle invariant as a whole. Similarly, an array of telephone poles, equally spaced along a straight line, has a translation symmetry—the entire array remains invariant if we perform a translation from any one pole to its neighbor.

The best way to acquaint oneself with these concepts is to trace a figure on a transparent sheet of paper, and then move the paper until the traced figure coincides with the original. For the equilateral triangle, there are six different ways in which this can be done, namely, three rotations of 120° each and three "flip overs" of the paper about the bisectors of the triangle, corresponding to the three reflections in these bisectors. Each of these operations will result in the traced figure overlapping the original. In a similar way, the array of telephone poles will coincide with itself after a translation, or sliding, of the transparent paper in the required amount and direction.

For convenience it is also customary to define a fourth symmetry operation, a *glide reflection;* this is the end result of a reflection in a given line, followed by a translation *along* that line (Fig. 20.4). A line of footsteps in the sand exhibits this kind of symmetry.[1]

---

[1] The four symmetry operations just described—reflections, rotations, translations, and glide reflections—have an important property in common: they preserve distance. That is, if P and Q are any two points in the plane, and if these points are carried over to P' and Q', then $PQ = P'Q'$. A distance-preserving transformation is called an *isometry* (from the Greek *isos* = equal and *metron* = measure). The four symmetry operations listed above comprise all possible isometries. (Of course, not every transformation is an isometry, as the example of inversion shows.) It is the distance-preserving property that makes the isometries so important in studying the symmetry of a figure; a symmetry operation leaves a figure invariant, and thus must preserve the distances between its points.

156

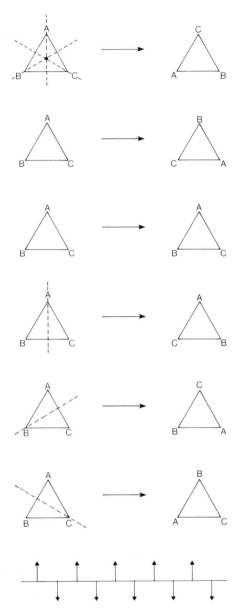

◁
Figure 20.4. *Glide reflection.*

While reflections and rotations alone suffice to study the symmetry of a single object such as a flower or a leaf, we need all four symmetry operations to analyze an *infinite pattern*. Such infinite patterns can be found almost everywhere in art and science; the band ornaments of ancient and primitive art and the wallpaper and floor designs of more recent times are but a few examples.

It is in the analysis of these infinite patterns that modern mathematics—and in particular a branch of algebra known as *group theory*[2]—has proven to be most effective.

Simplest of all infinite patterns is the one-dimensional *ornamental band*—a single figure (which itself may be two-dimensional) repeated regularly along an infinite strip.[3] It turns out that exactly seven different types of symmetry are possible here—the seven *one-dimensional symmetry groups*. They are listed in Table 3. In study-

Table 3. *The seven one-dimensional symmetry groups.*

	Typical pattern	Symmetry elements
1.	···L L L L···	1 translation
2.	···L Γ L Γ···	1 glide reflection
3.	···V V V V···	1 translation, 1 vertical reflection line
4.	···N N N N ···	1 translation, 1 half-turn*
5.	···V Λ V Λ···	1 vertical reflection line, 1 half-turn
6.	···E E E E···	1 translation, 1 horizontal reflection line
7.	···H H H H···	3 reflections (1 horizontal, 2 vertical)

* A half-turn is a 180° rotation.

ing this table, it is important to remember that the actual shape of the generating figure is irrelevant to the classification; only its symmetry elements (i.e., the different ways in which it can be repeated without changing the pattern as a whole) are important. For example, the *V*'s in pattern No. 3 could be replaced by *A*'s without affecting the type of symmetry of the band. They could also be replaced by the pair of letters *b* and *d,* generating the pattern . . . . *b d b d.* . . . That this pattern belongs to the same symmetry group as pattern No. 3 can be seen by placing a vertical mirror between any *b* and the *d* next to it; this, and the translation along the band, leave the pattern unchanged. Simi-

---

[2] The concept of a *group* is one of the most fundamental concepts of modern algebra. Introduced around 1830 as a purely abstract concept, it has found numerous applications in almost every branch of mathematics and science, as well as in art. A brief introduction to groups is found in the Appendix.

[3] One can, of course, think of an even simpler case—a one-dimensional "ornamental line" in which the generating figure is one-dimensional and confined to that line. Such a figure can only be a directed line segment, symbolized by an arrow. Just two types of symmetry are possible here, depending on whether two adjacent arrows point in the same or in opposite directions (Fig. 20.5). The first type is that of an endless convoy of identical cars moving in a straight line; the second type is the one-dimensional analog of the repeated reflections in a pair of parallel mirrors, as in the barbershop setting.

a

b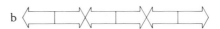

Figure 20.5. The two types of a one-dimensional ornamental line.

larly, the pattern . . . *b d p q b d p q* . . . belongs to the same group as pattern No. 5—it remains invariant when reflected in a vertical line placed between *b* and *d,* and also when turned upside down (a half-turn).[4]

That the seven symmetry groups listed above exhaust all possibilities of ornamenting an infinite band is rather remarkable, considering the unlimited variety from which we can choose the actual pattern. We have here a fascinating interplay between the freedom of our imagination to select the basic pattern and the restrictions imposed on it by the laws of geometry. It is perhaps this interplay, this "challenge of constraint," to quote E.H. Gombrich in *The Sense of Order,*[5] that has made the ornamental band so universally appealing. We can find it among the artifacts of almost every geographic and ethnic group and in every period from antiquity to the present. At the Heathrow underground station in London there is a large wall display of ornamental bands from different periods and cultures; you can view it while riding comfortably on a horizontal escalator that connects the airport terminal with the nearby subway station. It is an enticing preview to the endless treasures awaiting the visitor at the British Museum! Figure 20.6 shows a typical selection of ornamental bands, one from each

(*a*). Ancient Greek.

(*b*). Medieval.

(*c*). Ancient Greek.

(*d*). Navaho Indian.

(*e*). Islamic, sixteenth century.

(*f*). Inca, pre-Columbian.

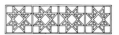

(*g*). Chinese, seventeenth century.

Figure 20.6. Examples of the seven types of one-dimensional ornamental bands, arranged according to their symmetry groups (see Table 3). Reprinted from Peter S. Stevens, Handbook of Regular Patterns: An Introduction to Symmetry in Two Dimensions, The MIT Press, 1981, with permission.

[4] One can actually generate the entire pattern by a repeated application of these operations. Thus, for the pattern . . . *b d b d* . . . , begin with *b,* then reflect it in a vertical line, obtaining the pair *b d.* By translating this pair along the band, the entire pattern is obtained. Similarly, the pattern . . . *b d p q b d p q* . . . can be obtained by placing a vertical mirror to the right of the *b,* again obtaining the pair *b d,* and then turning this pair, with its imaginary mirror, upside down, the center of rotation being to the right of the *d;* this will result in the pattern *b|d • p|q.* The process can now be repeated indefinitely, generating the entire band.

It must also be pointed out that the symmetry elements listed for each type in the table are not necessarily unique. For example, pattern No. 3 can also be described in terms of two reflections, one in the bisector of each *V,* the other in a vertical mirror midway between two adjacent *V*'s. That these two reflections result in the same symmetry group as that described in the table follows from a theorem about the combination of two reflections, discussed in the Appendix.

[5] See footnote 9.

159

Figure 20.7. *Two independent translations in the plane.*

symmetry group, arranged in the same order as in Table 3.

Not surprisingly, the number of symmetry possibilities increases as we turn from the one-dimensional ornamental band to the two-dimensional "ornamental plane," of which the most common examples are the various wallpaper designs or floor patterns repeated endlessly. The new feature here is the existence of *two* independent translations (rather than one), corresponding to the two dimensions of the plane. This is illustrated in Fig. 20.7, which shows the simplest way to ornament the plane; here the basic figure (the letter *b*) has no intrinsic symmetry, and the entire pattern is generated by translating this figure in two different directions. The positions occupied by the basic figure and its translations form the vertices of an *infinite lattice,* the framework of the entire pattern. This lattice is made up of infinitely many congruent parallelograms, each of which can be regarded as a "unit cell."[6] The whole structure resembles the lattice of atoms in a crystal, and indeed crystallography uses the same mathematical tools as we used here to analyze our two-dimensional "ornamental crystal."

Based on practical experience, it has been known for a long time that the number of symmetry types which can tessellate the plane is finite, but proof of this fact was accomplished only in 1891, when the Russian crystallographer E.S. Fedorov (1853–1919) enumerated exactly seventeen types. These seventeen *symmetry groups of the plane* exhaust all possibilities; they are analogous to the seven symmetry groups of the one-dimensional band.[7] That

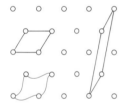

Figure 20.8. *Different unit cells for the same lattice.*

[6] The choice of the unit cell is not unique; different parallelograms may be chosen—or even non-parallelogram shapes—provided they do not contain any lattice points other than the vertices. (Figure 20.8 shows three possible choices.) It is this freedom to select one's basic shape that makes it possible to tessellate the entire plane with an artibrary design of the artist's choice, as in the work of M.C. Escher.

[7] As with the ornamental band, we must be careful not to confuse the seventeen symmetry groups with the infinite variety of the artistic design itself; the former is simply a classification of the pattern according to its symmetry elements, while the latter deals with the actual design that the artist chooses as his motif. The situation is somewhat similar to chemistry—there are only 92 natural elements, but from these an infinite number of compounds can be formed.

Also, the addition of color multiplies the symmetry possibilities many times, as can be seen from the simple example of the checkerboard, where just two colors create an entirely new pattern from the basic square lattice. This is the subject of polychromatic symmetry, a vastly more complex subject than the monochromatic symmetry discussed above. For a more technical introduction to this subject, see *Symmetry in Science and Art* by A.V. Shubnikov and V.A. Koptsik, translated from the Russian by G.D. Archard, Plenum Press, New York and London, 1974.

the number of different symmetry groups is not only finite, but indeed so small, is even more remarkable than is the case of the one-dimensional band; only the use of advanced methods from algebra (specifically, group theory) can provide an insight into the reasons for this. It is therefore all the more astonishing that examples of all seventeen groups have been found among the decorative designs of the ancients, and in particular the Egyptians. One might say that they discovered experimentally a fundamental mathematical fact whose proof was accomplished only some four thousand years later. Says Hermann Weyl (1885–1955) in his treatise, *Symmetry:*[8] "One can hardly overestimate the depth of geometric imagination and inventiveness reflected in these patterns . . . The art of ornament contains in implicit form the oldest piece of higher mathematics known to us." An example of each of the seventeen planar symmetry groups is shown in Fig. 20.9.

Of the many ethnic groups that have excelled in the art of ornamentation, none came closer to perfection than the Moslems. Barred by their religion from imitating the image of God—a decree which most Moslems take literally—they have devoted their artistic talents to creating abstract geometric designs of the utmost beauty. Everywhere in the Islamic world—from the Dome of the Rock in Jerusalem to the Alhambra in Granada—one stares with astonishment at the exquisite and intricate ornamental designs that decorate almost everything around: glazed ceramic tiles covering the walls of a mosque, huge oriental rugs on which the pious gather for their daily prayers, and brass plates and pottery artifacts of every shape and size. The motifs that decorate this opulence are as varied as the designs themselves, ranging from a host of floral themes—among which the arabesque is the most common— to purely geometric designs of stars and polygons, all interlaced in the most intricate manner. And to add to the magnificence, the designs are executed in a brilliant display of colors, dominated by blue and gold—the blue to symbolize the infinite and the gold to glorify the name of Allah.

The common theme of all Islamic art is geometric regularity, spatial rhythm, periodic repetition. Islam, with its central creed of an omnipotent God to whom all humans must humbly defer, found in the infinite pattern a supreme artistic expression of its philosophy. By showing only a finite portion of a design which in its entirety is infinite, the believer is reminded of his frailty and insignificance under the reign of the Almighty. The Moslems, who were assiduous students of Greek geometry, made here the

---

[8] Princeton University Press, Princeton, New Jersey, 1952.

Figure 20.9. *The seventeen ornamental symmetry groups of the plane. From Phares G. O'Daffer and Stanley R. Clemens,* Geometry: An Investigative Approach. *Copyright © 1976 by Addison-Wesley Publishing Company, Inc. Reprinted by permission.*

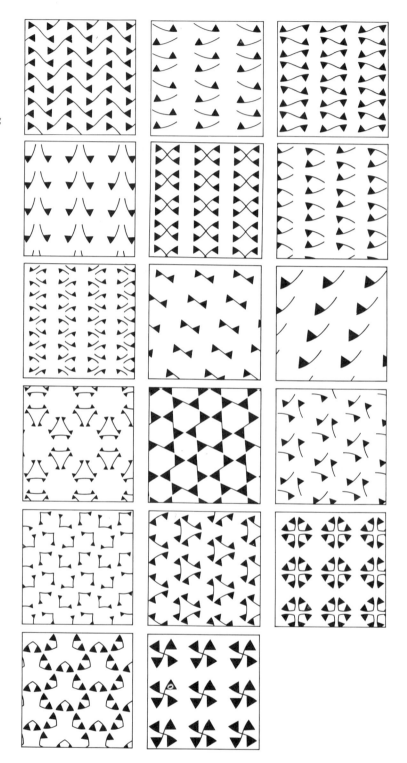

utmost artistic use of the mathematical heritage of their forebears—but with one difference: whereas the Greeks feared the infinite, the Moslems exulted in it, elevating it to the highest aesthetic and symbolic level. There are really two infinities at play here: the infinitely large, whose symbol is the regular repetition of a single motif, suggesting its continuation beyond its physical boundaries into infinity; and the infinitely small, as manifested by the desire to fill even the smallest empty space with some ornamental detail which then becomes part of the repetitive pattern. It is almost as though the Moslems feared empty space, replacing the Greek *horror infiniti* with a *horror vacui.* [9]

---

[9] I took the phrase "horror vacui, amor infiniti" from E.H. Gombrich, *The Sense of Order: A Study in the Psychology of Decorative Art,* Cornell University Press, Ithaca, New York, 1979. To quote Gombrich: "The urge which drives the decorator to go on filling any resultant void is generally described as *horror vacui,* which is supposedly characteristic of many non-classical styles. Maybe the term *amor infiniti,* the love of the infinite, would be a more fitting description."

21 Maurits C. Escher—
Master of the Infinite

*Deep, deep infinity! Quietness. To dream away from the tensions
of daily living; to sail over a calm sea at the prow of a ship, toward a
horizon that always recedes; to stare at the passing waves and listen
to their monotonous soft murmur; to dream away into
unconsciousness . . .*

— Maurits Cornelis Escher (1898–1972)[1]

*By keenly confronting the
enigmas that surround us,
and by considering and
analyzing the observations
that I had made, I ended
up in the domain of
mathematics. Although I
am absolutely without
training in the exact
sciences, I often seem to have
more in common with
mathematicians than with
my fellow-artists.*

You will not find his name in many art books, for he was largely
ignored by the art community. His pictures do not adorn the
walls of the world's great museums, for he loathed publicity. If
you wish to see his art, look for it in books on mathematics or
physics, for he felt a closer kinship to the world of science than
to his fellow artists. Unknown but to a few throughout most of
his lifetime, he suddenly rose to fame during the last fifteen years
of his life, but his genius was not universally recognized until
after his death. For if there has ever been an artist who depicted
the mathematical curiosities of the world around us, it was Maurits
Cornelis Escher.

Born in Leeuwarden, Holland, in 1898, Escher began his career
as a landscape painter. He particularly enjoyed the bright land-
scapes of the Mediterranean, and his colorful pictures of the little
towns and harbors of Italy and southern Spain have a charming
beauty and simplicity which is in sharp contrast to the highly intricate prints of his later years. These early works, in all likelihood,
would have ensured Escher a successful career as a landscape artist.
But it was during a visit to the Alhambra in Granada, in the
summer of 1936, that Escher's artistic interests took a complete
turn. He was profoundly impressed by the elaborate geometric
designs which decorate the walls of this great fourteenth-century
palace, a relic of the Moorish conquest of Spain. Escher spent

[1] All quotations in this chapter are Escher's. They are reprinted from
F.H. Bool, J.R. Kirst, J.L. Locher and F. Wierda: *M.C. Escher: His Life
and Complete Graphic Work,* (Harry N. Abrams, Inc.) 1981, and from
B. Ernst: *Magic Mirror of M.C. Escher* (Random House, Inc.) with permission.

164

Figure 21.1. *M.C. Escher: Waterfall* (*1961*). ©
M.C. Escher Heirs c/o Cordon Art—Baarn— Holland.

165

three full days there, diligently studying the variety of geometric motifs and copying some of them for later study. What he saw left a lasting impression on him; from then on, his work would become increasingly geometric in character. Years later he would reflect on his obsession with tessellations that was sparked by the Alhambra visit: "It is the richest source of inspiration that I have ever tapped, and it has by no means dried up yet."

Many mathematical concepts play a role in Escher's later work: infinity, relativity, reflections and inversions, and the relation between a three-dimensional object and its depiction on a two-dimensional surface. Above all, the notion of symmetry, in its broadest sense, is central in his work. All the four symmetry operations come into play here, to which Escher adds a fifth: similarity. This, and an endless fascination with infinity, is the essence of Escher's work.[2]

Escher's work relating to infinity can be divided into three categories:

1. Endless cycles.
2. The regular division of the plane.
3. Limits.

I don't want to be labelled as an artist. What I have always aimed at doing is to depict clearly defined things in the best possible way and with the greatest exactitude.

In the first category Escher has expressed his fascination with rhythm, regularity, and periodicity by depicting on two-dimensional canvas what has eluded generations of inventors and daydreamers in the real world: perpetual motion. Always using some subtle twist or hidden "trick," these prints have something of the grotesque about them, as though Escher enjoys poking fun at the laws of nature. In his own words: "I cannot help mocking all our unwavering certainties. It is, for example, great fun deliberately to confuse two and three dimensions, the plane and space, or to poke fun at gravity." We have already seen how he had used the topological peculiarities of the Möbius strip to depict a parade of horsemen or a bunch of giant ants chasing each other in an endless cycle. In *Waterfall* (1961; Fig. 21.1) he cleverly distorts the outline of a building to present us with an absurd situation: a water stream flowing endlessly in a closed circuit and causing a wheel to turn as the water plunges down—a machine feeding on its own energy. And in *Ascending and Descending* (1960; Fig. 21.2) Escher makes an ingenious use of the laws of perspective to show us a platoon of soldiers climbing up a stairway; up and

I have never attempted to depict anything mystic: what some people claim to be mysterious is nothing more than a conscious or unconscious deceit! I have played a lot of tricks, and I have had a fine old time expressing concepts in visual terms . . . All I am doing in my prints is to offer a report of my discoveries.

[2] In recent years an extensive literature has been published dealing with the mathematical aspects of Escher's work. Some of this literature is listed in the Bibliography.

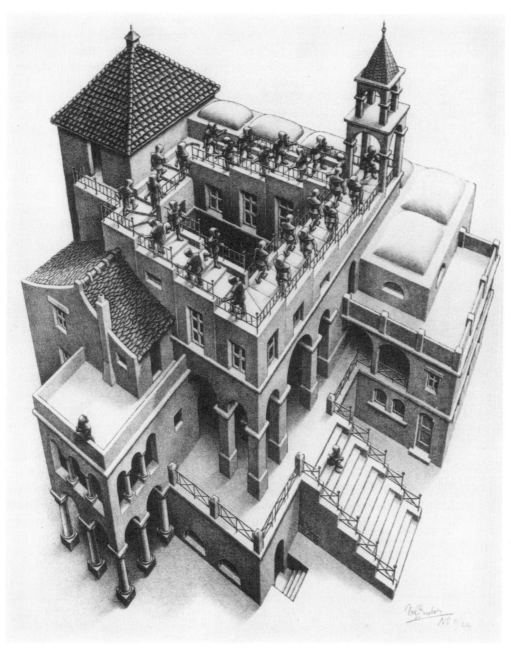

Figure 21.2. *M.C. Escher:*
Ascending and
Descending (1960). ©
M.C. Escher Heirs c/o
Cordon Art—Baarn—
Holland.

Escher's definition of tessellation: *A plane, which should be considered limitless on all sides, can be filled with or divided into similar geometric figures that border each other on all sides without leaving any 'empty spaces.' This can be carried on to infinity according to a limited number of systems.*

up they climb—only to find themselves back at the starting point! Said he: "Yes, yes, we climb up and up, we imagine we are ascending; every step is about ten inches high, terribly tiring—and where does it all get us? Nowhere; we don't get a step farther or higher."

It is the second category, the regular division of the plane (and in some cases, also of space), that has become Escher's hallmark. The possibility of repeating endlessly a single motif, without overlapping and without leaving any empty space, presented to him an irresistible challenge: "It remained an extremely absorbing activity, a real mania to which I have become addicted, and from which I sometimes find it hard to tear myself away." But unlike the Islamic designs which had so much inspired him, Escher's motifs are seldom abstract; on the contrary, they are recognizable objects—human beings, birds, fish, and inanimate objects taken from daily life. Escher expressed his distaste for the purely abstract in these words:

I feel as though it is not I who determines the design, but rather that the simple little plane compartments on which I fuss and labor have a will of their own, as though it is they that control the movements of my hand.

> *The Moors were masters in the filling of a surface with congruent figures . . . What a pity it was that Islam forbade the making of 'images.' In their tessellations they restricted themselves to figures with abstracted geometrical shapes . . . I find this restriction all the more unacceptable because it is the recognizability of the components of my own patterns that is the reason for my never-ceasing interest in this domain.*

It is Escher's ability to portray abstract mathematical ideas in terms of concrete, recognizable objects that is perhaps his greatest genius. Compare, for example, Figs. 21.3 and 21.4: the first shows a Greek abstract design from the sixth century B.C.; the second is by Escher. The two figures belong to exactly the same symmetry group—both permit just two translations, one along each row, the other across two rows.[3] While the Greek design, though aesthetically pleasing, is not particularly interesting, Escher's comes alive with a progression of identical Pegasi that fill the entire figure. On closer inspection we find that each black Pegasus is surrounded by four identical white Pegasi, and vice versa! In effect, the picture can be interpreted in two equally valid ways—black Pegasi flying against a white background or white ones against a black background. This illustrates another of Escher's favorite themes—duality. The dual effect is achieved here through an ingenious use of symmetry principles: the "empty" spaces between

---

[3] This is the same group that was discussed on p. 160. It is the simplest of all the seventeen planar symmetry groups, and is designated as group $p1$. The other groups mentioned in this chapter have, in their order of presentation, the designations pg, cm, and $p31m$.

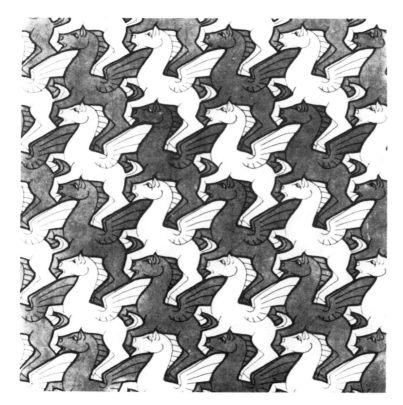

Figure 21.3. *Greek design, sixth century B.C. Reprinted from Peter S. Stevens,* Handbook of Regular Patterns: An Introduction to Symmetry in Two Dimensions, *The MIT Press, 1981, with permission.*

◁

Figure 21.4. *M.C. Escher: Pegasi.* © *M.C. Escher Heirs c/o Cordon Art— Baarn—Holland.*

adjacent Pegasi, whether white or black, are exact copies of the very same Pegasi, except that their color is reversed.

Next, let us compare Figs. 21.5 and 21.6. Again both belong to the same symmetry group, consisting of a translation along each horizontal band and a glide reflection in the border line between two adjacent vertical bands.[4] Again the first figure, a Peruvian design, depicts an abstract motif—or almost abstract, if one interprets the shape inside each frame as the head of a swan. But look what Escher does with the same basic pattern: he fills it with two arrays of horsemen, the black ones riding to the left, and the white ones—exact mirror copies of the black—riding to the right. And once more, each group of horsemen completely fills the space between the horses of the opposite group. This design is the same Escher had used eleven years earlier in his

---

[4] This group can also be described in terms of two glide reflections in the border lines between three adjacent vertical bands.

169

Figure 21.5. *Peruvian design. Reprinted from Peter S. Stevens,* Handbook of Regular Patterns: An Introduction to Symmetry in Two Dimensions, *The MIT Press, 1981, with permission.*

Figure 21.6. *M.C. Escher: Regular Division of the Plane III (1957).* © *M.C. Escher Heirs c/o Cordon Art—Baarn—Holland.*

Figure 21.7. *Symmetry elements for the designs in Figs. 21.8 through 21.11. Reprinted from Peter S. Stevens,* Handbook of Regular Patterns: An Introduction to Symmetry in Two Dimensions, *The MIT Press, 1981, with permission.*

Möbius strip print, *Horsemen* (1946); this time he simply called it *Regular Division of the Plane III*.

In the Pegasi and horsemen designs, the basic motif itself had no internal symmetry (and no wonder, since we are looking at the horses in profile). In the next five figures, the fundamental motif has a bilateral (reflection) symmetry, which puts the design in a new symmetry group; this group consists of a reflection and two glide reflections, all in parallel vertical lines (Fig. 21.7). Designs belonging to this group are very common—Figs. 21.8, 21.9, and 21.10 show, respectively, a medieval design based on a heraldic motif, a Japanese design from the nineteenth century, and a design from the Alhambra. Finally, Fig. 21.11 shows a portion

170

of Escher's *Regular Division of the Plane II,* a print which he designed for a book he wrote in 1957 with the same title.

In the previous designs, only translations, reflections, and glide reflections played a role. If one adds rotations, new symmetry groups are formed. Figure 21.12 shows an Arabic design typical of Islamic Persian architecture. Its symmetry elements are shown in Fig. 21.13, in which the thin lines indicate mirror lines and the little triangles represent 120°-rotation centers. Escher based one of his most beautiful prints on this pattern, which we reproduce here in color (Plate III, see p. 144).

Of the seventeen planar symmetry groups, at least thirteen are represented in Escher's work.[5] This shows that he must have had a deep, intuitive grasp of mathematical principles, and yet he had no formal mathematical training beyond high school. In his own words:

Figure 21.8. *Medieval heraldic motif. Reprinted from Peter S. Stevens,* Handbook of Regular Patterns: An Introduction to Symmetry in Two Dimensions, *The MIT Press, 1981, with permission.*

> *I never got a passing mark in mathematics. The funny thing is I seem to latch on to mathematical theories without realizing what is happening. No indeed, I was a pretty poor pupil at school. And just imagine—mathematicians now use my prints to illustrate their books! Fancy me consorting with all these learned folk, as though I were their long-lost brother. I guess they are quite unaware of the fact that I'm ignorant about the whole thing.*

Up until 1955, Escher used only congruent figures in his tessellations. But around that time he began to explore a new approach to infinity, an approach which ultimately would lead to some of his greatest masterpieces. What brought about this change we do not know; it was perhaps his feeling that he had already achieved a complete fulfillment with his previous tessellations (or, as he preferred to call them, regular divisions of the plane), and that nothing more could be accomplished in this realm. Or it may be that he was dissatisfied with the ability of tessellations to convey a true sense of the infinite. In an article published in 1959 he wrote:

Figure 21.9. *Japanese design, nineteenth century. Reprinted from Peter S. Stevens,* Handbook of Regular Patterns: An Introduction to Symmetry in Two Dimensions, *The MIT Press, 1981, with permission.*

> *What has been achieved in periodic surface division. . . ? Not infinity, of course, but certainly a fragment of it, a part of the 'reptilian universe.' If this surface, on which forms fit into one another, were to be of infinite size,*

---

[5] The classification of Escher's works according to the seventeen planar symmetry groups is made somewhat difficult by the fact that many of his prints use a two-color symmetry, as exemplified by Fig. 21.6: the figure remains invariant under a horizontal translation, but the vertical glide reflection results in a color reversal of the horsemen. (In strictly monochromatic symmetry only the shape remains invariant.) The classification becomes even more complex in his multi-color prints, as in Plate III. For a more thorough discussion of this subject, see *Fantasy & Symmetry: The Periodic Drawings of M.C. Escher,* by Caroline H. MacGillavry, Harry N. Abrams, New York, 1976.

Figure 21.10. *Design from the Alhambra, Granada, Spain. Reprinted from Peter S. Stevens,* Handbook of Regular Patterns: An Introduction to Symmetry in Two Dimensions, *The MIT Press, 1981, with permission.*

Figure 21.11. *M.C. Escher:* Regular Division of the Plane II *(1957).* © *M.C. Escher Heirs c/o Cordon Art—Baarn—Holland.*

Figure 21.12. *Arabic design. Reprinted from Peter S. Stevens,* Handbook of Regular Patterns: An Introduction to Symmetry in Two Dimensions, *The MIT Press, 1981, with permission.*

then an infinite number of them could be shown upon it. But we are not simply playing a mental game: we are conscious of living in a material, three-dimensional reality, and it is quite beyond the bounds of possibility to fabricate a flat surface stretching endlessly and in all directions.

Escher's solution to this problem was to add a fifth symmetry operation to the four previous ones: similarity. By relaxing the requirement that a figure should retain its shape *and* size, and insisting instead on invariance of shape alone, it is possible to convey the suggestion of infinity without actually reaching it. This is the principle which, from then on, would dominate Escher's work.

The first of these "limit prints," *Smaller and Smaller I* (1956), is shown in Plate IV on p. 145. We see a whirlwind of reptiles, all having exactly the same shape (allowance being given for mirror reversals), but diminishing in size as we approach the center. In fact, Escher lets the size of his reptiles follow the geometric progression 1, ½, ¼, ⅛, . . . , as is clear from the grid he had used for this print (Fig. 21.14). Thus the center becomes the point "where the limit of the infinitely many and infinitely small is reached." But he was not quite pleased with the result, for "the reduction from without inwards" did not convey his "longing for an intact and complete symbol of infinity." What he was looking for was a method for "reduction from within outwards," and he found it in an illustration which he saw in a book by the Canadian mathematician H.S.M. Coxeter. Coxeter's illustration (Fig. 21.15) was in connection with Henri Poincaré's model of non-Euclidean geometry (see p. 127); Escher, however, did not care much about its theoretical significance, but immediately realized its aesthetic potential:

I am trying to glean from it a method for reducing a plane-filling motif which goes from the centre of a circle out to the edge, where the motifs will be infinitely close together. His [Coxeter's] hocus-pocus text is no use to me at all, but the picture can probably help me to produce a division of the plane which promises to become an entirely new variation of my series of divisions of the plane. A circular regular division of the plane, logically bordered on all sides by the infinitesimal, is something truly beautiful . . .

He added wryly: "At the same time I get the feeling that I am moving farther and farther away from work that would be a 'success' with the 'public,' but what can I do when this sort of problem fascinates me so much that I cannot leave it alone?"

From Coxeter's illustration there evolved four of Escher's most successful prints; he called them simply "Circle Limits." In the first of these, *Circle Limit I* (1958; Fig. 21.16), Escher used alternating white and black winged fish that move along arcs of circles

172

perpendicular to the boundary circle. But the fact that fish of the same color always meet either head-to-head or tail-to-tail left much to be desired, for, in Escher's words, "there is no continuity, no 'traffic flow,' nor unity of color in each row." These shortcomings he remedied in *Circle Limit III* (1959, Plate V, see p. 146), the most celebrated of his limit prints. Here is Escher's own description of this work:

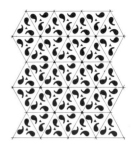

> In the colored woodcut Circle Limit III the shortcomings of Circle Limit I are largely eliminated. We now have none but 'through traffic' series, and all the fish belonging to one series have the same color and swim after each other head to tail along a circular route from edge to edge. The nearer they get to the center the larger they become. Four colors are needed so that each row can be in complete contrast to its surroundings. As all these strings of fish shoot up like rockets from the infinite distance at right angles from the boundary and fall back again whence they came, not one single component ever reaches the edge. For beyond that there is 'absolute nothingness.' And yet this round world cannot exist without the emptiness around it, not simply because 'within' presupposes 'without,' but also because it is out there in the 'nothingness' that the center points of the arcs that go to build up the framework are fixed with such geometric exactitude.

Figure 21.13. *Symmetry elements for Fig. 21.12 and Plate III. Reprinted from Peter S. Stevens,* Handbook of Regular Patterns: An Introduction to Symmetry in Two Dimensions, *The MIT Press, 1981, with permission.*

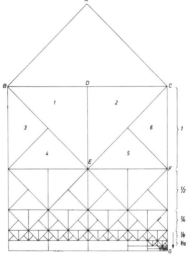

Figure 21.14. *M.C. Escher: grid for* Smaller and Smaller. © *M.C. Escher Heirs c/o Cordon Art—Baarn—Holland.*

Could anyone have conveyed the essence of Poincaré's model more succinctly?[6]

---

[6] Escher sent a copy of *Circle Limit III* to Coxeter, whose reply bemused him: "I had an enthusiastic letter from Coxeter about my coloured fish, which I sent him. Three pages of explanations of what I actually did . . . It's a pity that I understand nothing, absolutely nothing of it . . ." Coxeter once took Escher to one of his lectures on non-Euclidean geometry, convinced that he would be able to follow the subject. As might be surmised from Escher's remarks, Coxeter's efforts remained unfulfilled.

Figure 21.15. *The Coxeter illustration. From* H.S.M. Coxeter, Introduction to Geometry. *Copyright ©* *1969 by John Wiley and Sons, Inc. Reprinted by permission of John Wiley and Sons, Inc.*

Figure 21.16. *M.C. Escher:* Circle Limit I *(1958). © M.C. Escher Heirs c/o Cordon Art— Baarn—Holland.*

One other series of "limit prints" should be mentioned—the series based on the logarithmic spiral. As we have seen, this curve has the unique property that every segment of it is similar in shape to any other segment (see p. 75). It is no wonder, then, that Escher found in the logarithmic spiral a most fitting motif with which to express his ideas. In *Path of Life II* (1958; Fig. 21.17) he used four interwoven spirals along which grey and white fish move inward and outward; on closer inspection we see that the inward-moving fish are all grey, while the outward-moving ones are all white. There is, of course, a hidden symbolism

Figure 21.17. *M.C. Escher:* Path of Life II (1958). © *M.C. Escher Heirs c/o Cordon Art— Baarn—Holland.*

here: the white fish, emanating from the center, represent birth and growth; upon reaching the edge their color turns into grey, and they sink back to the center from which they emerged, completing the cycle of life and death.

The same idea found its supreme expression in *Whirlpools* (1957; Plate VI, see p. 147), which in my opinion is the most beautiful of all of Escher's works. Two systems of logarithmic spirals, running parallel to each other, emanate from the top center, and after an infinite number of revolutions attain their maximum size at the center of the print; then they diminish again until they reach the lower center. Along these spirals, red and grey fish swim peacefully—the red fish originating from the lower center and moving towards the upper, the grey ones in the opposite direction. The entire picture can be turned through 180° about its center—viewing it upside down will merely turn the red fish into grey ones and conversely. (Even Escher's signature appears twice—in the lower right corner and again upside down in the upper left.) Most of Escher's ideas are embodied here in the most

On his print, *Whirlpools* (1957): *I designed a division of the plane consisting solely of fish that 'move' towards the centre (symbolizing death or dying), while a series of white fish 'move' outwards from the same centre (life, birth). The attractive, and at the same time difficult, thing is the diminution of the fish figures into infinity. The outer fish will be about five feet long and I want to try and reduce their size consistently until they are mere specks of about half an inch in length.*
(Comments made by Escher regarding the commission he received for a large mural based on this print.)

175

masterful way—his fascination with the infinite, for the two centers are infinitely remote and thus forever inaccessible to our fish; his lifelong obsession with tessellations, congruence, and similarity (not only does each red fish have an exact counterpart among the grey ones, but the fish completely fill the space around and between the spirals); and finally, his extraordinary talent for depicting movement, change, cycle, and rhythm. It is perhaps symbolic that Escher was commissioned by the city of Utrecht to use the same design in a mural for one of the city's cemeteries. He himself painted this large mural, 3.7 meters in diameter.

Figure 21.18. *J.S. Bach: Crab canon from* The Musical Offering *(1747)*.

And so we are reminded again of the mathematical insight of an artist who has hardly had any mathematical training at all. In this Escher resembles another artist who, though living two centuries before him and working in an entirely different medium, nevertheless shared with him a similar mathematical intuition: Johann Sebastian Bach (1685–1750). Both men despised publicity, and both were fully recognized only after their death. Above all, both had an acute sense for pattern, rhythm, and regularity—temporal regularity in Bach's case, spatial in Escher's. Though neither would admit it (or even be aware of it), both were experimental mathematicians of the highest rank.[7] Perhaps it is no coincidence that Bach's music and Escher's drawings have in our time become the favorite subject of popular art of every kind—Bach's melodies being transcribed to anything from "pop" and rock music to computerized and synthesized sound, and Escher's motifs bla-

Again on Whirlpools: *I am using a new printing technique based on a very amusing twofold-rotation-point system. It is difficult to explain in words, but what it amounts to is that from each of the blocks (probably three) that I have to cut, I make only half of the surface that they have to fill together; the other half is produced by repeating the blocks after they have been turned a hundred and eighty degrees. I doubt whether 'the public' will ever understand, let alone appreciate, what fascinating mental gymnastics are required to compose this sort of print.*

[7] The notion of symmetry is as central to Bach's music as it is to Escher's drawings. The well-tempered scale, which due to Bach's efforts has become the standard scale of Western music, is actually based on a symmetry group. (Since all twelve tones of this scale are musically equivalent, a

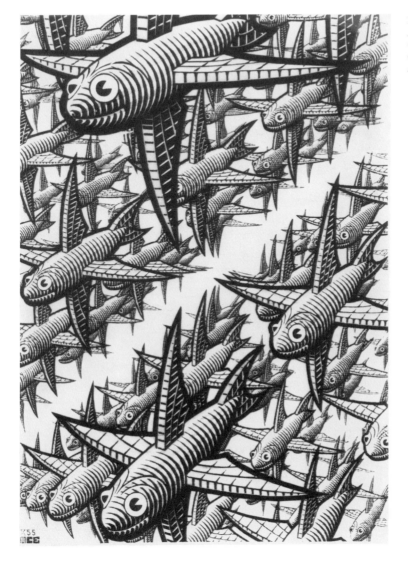

Figure 21.19. *M.C. Escher:* Depth (1955). © *M.C. Escher Heirs c/o Cordon Art—Baarn—Holland.*

melody played in any one key will remain invariant when transposed to any other key.) Bach often made explicit use of symmetry principles in his works; perhaps the most famous example is the "crab canon" from *The Musical Offering* (1747; Fig. 21.18), in which the melody is also its own accompaniment in retrograde motion—each is the other's exact mirror image in time. This can be regarded as the musical equivalent of Escher's *Horsemen,* where subject and background are reverses of each other in space. The crab canon is the opening theme in Douglas R. Hofstadter's book *Gödel, Escher, Bach: an Eternal Golden Braid,* Basic Books, New York, 1979.

On his *Smaller and Smaller* (1956): *I went to some trouble to explain the disk-shaped print* Smaller and Smaller *to a number of visitors, but it's becoming increasingly clear that on the whole people are not sensitive to the beauty of this infinite world-in-an-enclosed-plane. Most people simply do not understand what it is all about.*

Figure 21.20. *M.C. Escher:* Cubic Space Division (*1952*). © *M.C. Escher Heirs c/o Cordon Art—Baarn—Holland.*

Above all I am happy about the contact and friendship of mathematicians that resulted from it all. They have often given me new ideas, and sometimes there even is an interaction between us. How playful they can be, those learned ladies and gentlemen!

zoning from T-shirts, posters, and record and book covers. Whether either would have welcomed this craze is doubtful, but in this age of space travel and computers, it is perhaps a final tribute to two minds who, each in his own way, have united the paths of art and science.

178

The Modern Kabbalists 22

> . . . *Here*
> *In the interminable wilderness*
> *Of worlds, at whose immensity*
> *Even soaring fancy staggers.*
>
> — Percy Bysshe Shelley
> (1792–1822)

During the twelfth century there evolved in central Europe a mystic movement of Jewish devotees, the kabbalists, whose belief in the transcendence of God led them to the *Ein Sof,* the infinite. According to the *Kabbalah* (in Hebrew: "tradition"), God is revealed to man only through His virtues and deeds, never directly as Himself; the many references to God in the scriptures are only allusions to His manifestations. In their search for spiritual fulfillment, the kabbalists were seeking a path to the divine spirit, if not to God Himself. This they achieved through a system of ten *sephirot* (literally: "spheres," and also "enumerations"), emanating from the *Ein Sof* ("endless"), which became the symbol for the hidden God (Fig. 22.1). The upper sphere, the one closest to the *Ein Sof,* was called the "crown"; next came the spheres of "wisdom," "intelligence," "mercy," and so on down to the lowest sphere, the "kingdom." It is only through the ten sephirot, according to the kabbalists, that one can approach the divine spirit; perhaps we can find here a subtle reference to the mathematical idea of limit, of an infinite series whose sum we can only approach, never reach. The kabbalists depicted their sephirot in various geometric forms, as in Fig. 22.1. The system gradually increased in complexity, with new interpretations being added by subsequent generations. The Kabbalah gradually spread throughout Europe and reached its climax in Spain prior to the expulsion of the Jews from that country in 1492. After that it moved to Poland and to Jerusalem, where it evolved into the modern Hasidic movement.

Infinities and indivisibles transcend our finite understanding, the former on account of their magnitude, the latter because of their smallness; Imagine what they are when combined.
□ Galileo Galilei (1564–1642), statement made by Salviati in *Dialogues of Two New Sciences* (1638)

179

Figure 22.1. *The ten* sephirot *of the Kabbalah, emanating from the* Ein Sof *("the infinite"). The inscriptions are in Hebrew. Reprinted from Z'ev ben Shimon Halevi,* Kabbalah: Tradition of Hidden Knowledge, *Thames and Hudon, 1979, with permission of the author. Illustration by James Russell.*

Judaism, which has never revered abstract symbols, let alone material artifacts, nevertheless found in the *Ein Sof,* the infinite, a symbolic expression for its aspirations to the divine spirit.[1] Such aspirations, of course, are common to all religions. Islam, with its central philosophy of an omnipotent God who reigns the universe, expressed its religious yearnings in the exquisitely beautiful and complex geometric designs of which the Moslems were masters. And the oriental religions, Hinduism and Buddhism, revered another kind of infinity, eternity, through their belief in the reincarnation of the soul in endless cycles.

But without question, it was Christianity that gave the idea of infinity its most visible manifestation. The theme of the resurrection, so central to the Christian faith, found its ultimate expression in the vast churches and cathedrals of Gothic, and later Baroque, Europe. The church, once a place for prayer and solitude, was gradually transformed into an immense shrine to celebrate the glory of God. To stand up to this task, the Gothic architects planned and built huge, awe-inspiring structures in which the believer would feel the presence of divinity. Dwarfed by their im-

---

[1] The so-called tetragrammaton, the four Hebrew letters יהוה (usually transliterated into YHWH) that form the Biblical proper name of God, is a fusion of the three Hebrew words היה ("He was"), הווה ("He is"), and יהיה ("He will be"), signifying the eternity of God's presence.

mense dimensions, the visitor to these shrines was overwhelmed not once, but twice: first from the outside, by their sheer heights; and, once inside, by the vast spaces of their interiors. The sky-soaring heights of the Gothic cathedral, with its "spires whose silent finger points to heaven" (to quote William Wordsworth in *The Excursion*), gave the illusion that the entire structure was defying the laws of gravity, lifting itself from its earthbound foundations to soar to infinity. And its dimly lit interior, pierced here and there by shafts of light penetrating through a window high above the floor, created in the visitor a vision of the immensity of God's creation.

The desire to achieve a sense of inspiration and awe through structures of vast dimensions reached its climax during the Baroque period. Marked by the great discoveries of Copernicus, Galileo, Kepler, and Newton, the new era signaled a vast expansion of the universe—the physical as well as the spiritual. The earth, no longer the center of the universe, became but a tiny speck in an infinite cosmos; and the mind, having been liberated from long-entrenched religious dogmas, was free to seek new and unexplored vistas. All this brought about a triumph of the infinite. Once barred from science and assailed by the Greeks as a "horror," it now became the central theme of a new era. The infinitely large at the one end of the scale, and the infinitely small, the infinitesimal, at the other, combined to usher in a new and dynamic conception of our world—as opposed to the static, finite universe of the Greeks. Far from contradicting each other, the two infinities actually combined forces in a most fruitful way. The newly invented differential and integral calculus, with the infinitesimal as its cornerstone, enabled physicists to explain terrestrial and astronomical events on the grandest scale—from the phenomenon of tides to the motion of the planets. Combined with Newton's universal law of gravitation, it gave rise to a grand, unified conception of the universe, one based on mathematical reasoning rather than theological speculation.

The consequences of the new world outlook were profound, and far transcended the realm of science. Art, architecture, and music all underwent an expansion on an unprecedented scale. No longer confined to the salons of the rich, the treasures of art now became accessible to the bourgeois, and larger halls were needed to accommodate them. The great Gothic cathedrals, in addition to performing their religious functions, became vast museums of art, and new shrines were built with this dual role in mind. It is hardly a coincidence that Sir Christopher Wren (1632–1723), the famed London architect, started work on his master-

As it was in the beginning, so it is now, and ever shall be, world without end.
□ From the Gloria of the *Magnificat*, Johann Sebastian Bach (1685–1750)

The Lord reigneth, the Lord hath reigned, the Lord will reign for ever and ever.
□ From the prayer recited before taking out the Torah

181

One generation passeth away, and another generation cometh: but the earth abideth for ever.
□ Ecclesiastes 1:4

piece, St. Paul's Cathedral, in 1675—within ten years of Newton's discovery of the calculus—and completed it in 1711, the very same year in which the calculus was first published. (Newton's *Principia,* in which he enunciated his universal law of gravitation, was also published during this period, in 1687.) Everything in this monumental structure—from its great dome to the colossal dimensions of its interior—was an architectural celebration of the new universe.

In music, too, the new era marked the beginning of a new style, with an enormous expansion of the size and tonal range of the orchestra. Musical performances, once the privilege of the rich, now became open to the public, who enjoyed them thoroughly. We can well imagine the effect that Handel's *Water Music,* or his *Music for the Royal Fireworks,* must have had on the London crowds gathered along the Thames; before them, floating down the river on barges, was a huge orchestra that filled the air with sonorities and harmonies never heard before, accompanied by a display of fireworks which lit the skies in an unforgettable spectacle of exuberance. The Baroque period, with its grandiose style, was the precursor of Classical and later Romantic music, culminating in the immense works of Berlioz, Wagner, and Mahler. Gustav Mahler's Eighth Symphony (1906), known as the "Symphony of the Thousand," reached gigantic proportions and required no fewer than eight soloists, a double choir, and a huge orchestra— a combined force ten times as large as the Mozartean ensemble of a mere century earlier.

It seemed, then, towards the end of the nineteenth century, that the infinite was finally liberated from the "horror" once attached to it, and was celebrating its triumph on a grand scale fitting its lofty meaning. But nothing in history ever remains final. As we now turn to the cosmological aspects of infinity, we shall see that the twentieth century has witnessed, in more ways than one, a diminution of this triumph and a revival of the finite universe—although in a way which neither the ancient Greeks nor the Newtonian world could have foreseen.

There has been an earth for a little more than a billion years. As for the question of the end of it I advise: Wait and see.
□ Albert Einstein (1879–1955), in answer to a question from a child

Cosmological Infinity   Part IV

*The history of
astronomy is a history
of receding horizons.*

— Edwin Hubble
(1889–1953)

*Man staring at infinity
(attributed to Camille
Flammarion, nineteenth
century).*

23 The Ancient World

Twinkle, twinkle, little star
How I wonder what you are.

— A popular rhyme

The heavens declare the
glory of God;
And the firmamant showeth
His handiwork.
□ Psalms 19:1

From the dawn of recorded history, man has watched the skies above him, marveling at their mysteries and wondering about the myriads of stars that seem to be embedded, like tiny gems, in the celestial dome. What are those stars made of? How far are they? What message do they have for us? Questions such as these, inspired by man's awe at the grandeur of the creation, were the first step in the creation of a science of the heavens, astronomy. It is a paradoxical fact that astronomy, which studies the farthest objects we can think of, was the first discipline of knowledge to become a full-fledged science in the modern sense of the word. Compare this with geology or biology—the disciplines concerned with our planet and its inhabitants—which emerged as true sciences only in the past few centuries. It seems that the farther a mystery lies, the greater the urge to solve it!

Dominating man's speculations about the universe was one question: Is this universe finite or infinite? Does it have a boundary, and if so, how far is it? Or is it boundless, extending forever in every direction? Either possibility raises serious questions that seem to challenge our most fundamental notions about space and time. If the universe has a boundary, what is there beyond this boundary? Empty space? Nothingness? It is difficult indeed to think that if we go far enough in a given direction, we will reach a point beyond which nothing exists, not even space itself. But equally disturbing is the thought of an infinite universe, one that extends endlessly in space and time. What significance would there be for man in such a universe? Would it not deprive him of his (admittedly self-proclaimed) central role in God's creation?

184

The entire history of astronomy has been an endless struggle between these opposing views. The answers given and the "models" proposed have shifted back and forth from one extreme to the other, influenced by the scientific attitudes and, even more so, by the religious doctrines prevailing at the time. And as we shall see, the mystery has by no means been solved yet.

*   *   *

Astronomy was thriving in Babylon, in Egypt, in India and China, and in Central America as early as the second millenium B.C. Clay tablets have been found on which the Babylonians recorded, in cuneiform script, detailed observations of the sun and moon, of eclipses, and of the motion of the then five known planets. Egyptian and Mayan temples have been built according to strict astronomical specifications. The famous temple of Amon-Ra, the sun-god, at Karnak (Thebes) in upper Egypt, was planned so that once a year—on the day of the summer solstice—its inner sanctuary would be illuminated by the sun, while on all other days it was shrouded in darkness. This event—and a spectacular one it must have been—had a significance far greater than its astronomical origin might suggest, for it coincided with the yearly rise of the Nile, upon which the very existence of Egypt depended. Above all, the ancients excelled in the art of timekeeping, and one cannot fail to marvel at the accuracy of the Babylonian and Mayan calendars. True, their interest in astronomy was motivated as much by religious and mythological beliefs as by practical needs (such as the prediction of the seasons for agricultural purposes); but whatever their reasons, the ancients were keen observers of the sky and first-rate practitioners of astronomy.

We never cease to stand like curious children before the great Mystery into which we are born.
□ Albert Einstein (1879–1955)

It was, however, the Greeks who transformed astronomy from a practical art into an intellectual discipline, a science. In this they followed the same tradition they had established in mathematics—the insistence that every theory should be supported by a rational argument. The Greeks were the first to speculate about the physical nature of the world, and they supported their speculations with the observational evidence they had collected. Moreover, the Greeks were as much interested in questions pertaining to the universe as a whole as in the more mundane aspects of astronomy, and in this sense they should be credited with founding the science of cosmology.

The earliest Greek models of the universe were as yet primitive creations, based more on mythological beliefs than on hard evidence. Thales of Miletus (*ca.* 624–548 B.C.), one of the early philosophers of Ionia (now western Turkey), imagined the earth

185

to be a flat disk floating in a vast ocean and surrounded by an atmosphere of vapor. This was in accordance with his belief that water constitutes the "primary substance" from which everything else is made. The heavenly bodies—the sun, the moon, the planets, and the fixed stars—were floating in this atmosphere, apparently at a fixed distance above the earth, turning about it once in 24 hours. It was a crude model indeed, all the more so since Thales was one of the great minds of his time; he was well versed in mathematics, and is said to have predicted the total eclipse of May 28, 585 B.C., which occurred while the armies of Lydia (in Asia Minor) and Persia were engaged in a raging battle. So frightened were the warring armies by the sudden fall of darkness that they laid down their weapons and signed a peace treaty then and there.

Anaximander (ca. 610–546 B.C.), a pupil of Thales, refined Thales' model by replacing the disk-shaped earth with a cylinder; more importantly, he perceived the heavenly bodies as moving in distinct "shells," thus placing them at different distances from the earth. This was a major innovation, but strangely enough he put the stars closer to the earth than the moon. The phenomenon of occultations—the occasional disappearance of a star behind the moon due to the latter's motion—should have told him of his error, but for some reason he disregarded this simple evidence.

The idea of shells, i.e., spheres in which the heavenly bodies are embedded, became a permanent fixture in virtually all subsequent Greek models of the universe. Only the details differed, and subsequent astronomers mainly preoccupied themselves with the details of the mechanism of such a system. The main problem was to account for the complicated observed motion of the planets, and especially their occasional retrograde motion. (During retrograde motion, a planet seems to move from east to west, rather than the usual west-to-east motion.) To account for this complex phenomenon, more and more shells were proposed, culminating with the model of Aristotle (384–322 B.C.), which had no fewer than 56 spheres, all set in motion by the outermost "divine sphere." Whether the Greeks actually believed in the physical existence of these spheres, or whether the sphere model merely served as a convenient way to explain the observed motion of the planets (in the same way as the Bohr model of the atom, which was also based on the shell idea, explained the observed spectral lines of hydrogen), is hard to say. The important fact is that the Greeks were the first to propose a picture of the universe which, crude as it was, did account for the astronomical facts as they were then known.

It is not so much the shell model that should appear to us naive, but the vastly underestimated dimensions of the world as the ancients perceived it. Anaximander's earth, not surprisingly, extended from the western end of the Mediterranean at Gibraltar to the shores of the Indian Ocean, which spanned the then known world. As for the size of the heavenly bodies, the Greeks differed greatly in their estimates. Heraclitus (ca. 540–475 B.C.) imagined the sun to be a fiery disk one foot in diameter—an absurd figure even for his time—while only two centuries later Aristarchus (ca. 320–250 B.C.) put the figure at about seven earth diameters, or roughly 50,000 miles—still short of the true value by a factor of 16. Even more naive were the Greek estimates of the size of the universe. Empedocles (ca. 490–430 B.C.) put the crystal sphere enclosing the universe at three times the earth–moon distance, while later Greek philosophers refrained altogether from estimating the limits of the world. And well they did, for even the most daring Greek speculations could not have anticipated just how vast the universe is.

To the extent that observational evidence should be the basis for any scientific theory, we must give the Greeks at least some credit for arriving at such grossly underestimated figures. The phenomenon of *parallax*—the apparent shift in the position of an object as we change our own position—was well known to the Greeks and played a crucial role in their reasoning. While the sun, the moon, and the planets change their positions in a regular and predictable manner, there seems to be no change whatsoever in the positions of the fixed stars (which is, of course, why they are called "fixed"). For thousands of years, these stars have remained virtually unchanged in their constellations, giving the ancients the one permanent feature they needed to reassure their existence in a changing world. The lack of any observable parallax could be explained in two different ways: by assuming that the earth is stationary and located at the center of a finite universe, or by assuming that the stars are so distant that any parallax caused by the earth's motion would be far too small to be detected by our eyes. The Greeks chose the first explanation. To their finite mind—we are reminded again of their "horror infiniti"—it was inconceivable to think of a vast universe in which the earth would shrink to an insignificant point. And to all practical purposes the earth does, indeed, seem to be fixed motionlessly at the center of the celestial dome, disturbed there not even by the slightest jolt which might have revealed its motion. Thus the Greeks settled for the easier and more comfortable of the two possibilities: a stationary earth that is forever fixed at the center of a finite universe, whose boundary is the crystal sphere of the fixed stars.

Occasionally there were voices who dissented from this comfortable world picture. Democritus (*ca.* 470–380 B.C.), the founder of the atomistic school, had already speculated that the Milky Way may be a vast cluster of tiny stars, rather than the continuous band of diffuse light that reveals itself to the naked eye. This, of course, fitted well with his atomistic philosophy—that everything in the universe is made of a vast number of tiny indivisible atoms. But the implications went further than that, for if the Milky Way could be composed of an enormous number of stars, then perhaps its distance, too, might be enormous; this could explain at once why the unaided eye is unable to resolve it into its constituent stars. Aristotle was even more explicit: he declared that "the bulk of the earth is infinitesimal in comparison to the whole world that surrounds it." And there was one voice who suggested the most daring idea of all: that the sun, and not the earth, is at the center of the universe. This voice was Aristarchus of Samos (*ca.* 320–250 B.C.), who correctly interpreted the absence of an observable parallax of the fixed stars as evidence of a vast, virtually infinite universe. But as happened so often in the history of mankind, his idea came too early: the Greek mind simply could not conceive an infinite universe, one in which the earth is removed from its central position. The credit for discovering the heliocentric (= sun-centered) system had thus to wait for Copernicus.

To be sure, the Greek astronomers did not confine their attention to cosmological speculations alone. In matters closer to home they made some very significant discoveries. They were the first to recognize that the earth is spherical, and the first to actually measure its dimensions. Eratosthenes' celebrated measurement of the earth's circumference, a feat which he achieved in the year 240 B.C., came within 100 miles of the correct value. Evidently this value, about 24,900 miles, was not favored by his fellow Greeks, for it implied an earth vastly larger than the one they had known. But once the true dimensions of the earth were accepted, they became a yardstick by means of which astronomical distances could be expressed. For example, by observing the earth's shadow as it is cast on the moon during a total eclipse, Aristarchus was able to calculate the earth–moon distance to be about 40 earth diameters. This value was later corrected by Hipparchus (*ca.* 190–120 B.C.) to 30—very close to the true figure. In his calculations Hipparchus made use of the new science of trigonometry (literally: "the measurement of three angles," i.e., a triangle), which he himself had developed and which gave the

Greek astronomers an invaluable tool with which to estimate astronomical distances. "The science of trigonometry," said the historian Stanley L. Jaki, "was in a sense a precursor of the telescope. It brought faraway objects within the compass of measurement and first made it possible for man to penetrate in a quantitative manner the far reaches of space, with the result that the accepted ideas on the structure of the cosmos had to be drastically revised."[1]

But the "drastic revisions" had to wait for a while—for nearly 1,500 years, to be exact. In the second century A.D., the astronomer–geographer Claudius Ptolemaeus, better known as Ptolemy, summarized the Greek world picture as it was then accepted in his monumental work *Almagest*, a thirteen-volume compilation of astronomical knowledge comparable to Euclid's *Elements*.[2] This world picture was a simple one: a spherical earth permanently fixed at the center of a finite universe, whose boundary is the celestial dome of the fixed stars. Ptolemy's geocentric and finite universe was to be the rock foundation of European astronomy for the next fifteen centuries. What is more, it became the official dictum of the Roman Catholic church, to be followed unequivocally by all its believers. Any deviation from this dictum was to be regarded as heresy, any show of independent thought ruthlessly suppressed. The consequences were devastating to the advent of science. Europe was about to plunge into the Dark Ages.

---

[1] *The Relevance of Physics,* The University of Chicago Press, 1966.

[2] As with Euclid four centuries before him, not much is known about Ptolemy, not even his years of birth and death. (He was unrelated to the royal family of the Ptolemies who had ruled Egypt half a century before him.) Both men lived in Alexandria, the intellectual center of the ancient world, and it was there that they wrote their works. Like Euclid's *Elements* (also in thirteen volumes), the *Almagest* does not contain Ptolemy's own discoveries but is a compilation of the astronomical observations and theories of his predecessors, including a catalog of some 1,000 stars based on Hipparchus's work. In this catalog Ptolemy listed and gave names to 48 constellations of stars, and these names are the ones still used today.

The name *Almagest* is Arabic, meaning "the greatest." It is an adoption of the earlier name *Syntaxis Mathematica* ("mathematical collection"), to which later generations added the superlative *magiste* ("greatest"), to distinguish it from other, less important works. Like most of the Greek works, Ptolemy's book became known to the Western world through its Arabic translation; thus the Greek *magiste* became the Arabic *Almagest.* The first Latin translation appeared in 1175, and from then on until the sixteenth century it dominated the astronomical thinking of Europe.

189

24 The New Cosmology

Open the door through which we may look into the limitless, unified firmament!

— Giordano Bruno (1548–1600)

It was not that astronomy came to a complete standstill during the Middle Ages. Many Arab and Jewish astronomers, working largely in Spain under the Islamic conquest but also in Persia and Turkey, made extensive observations of the stars and planets and used these observations to refine astronomical tables and almanacs. Even more significantly, these scholars rediscovered many of the Greek works in mathematics and astronomy and translated them into Arabic and thence into Latin. It is mainly through these translations that our knowledge of Greek science became possible. But important as these contributions were, they did not change man's fundamental picture of the universe. This picture was essentially the Aristotelian–Ptolemaic one, according to which the immovable earth is at the center of a finite universe, made up of spherical shells in which the planets and stars are embedded.

The changes about to come in this world picture had a slow beginning. Nicolaus of Cusa (or Cusanus, 1401–1464), a German scholar and theologian who studied at the University of Padua in Italy, was perhaps the first to contemplate an infinite universe. In his best-known work, *De docta ignorantia* ("On Learned Ignorance," published posthumously in 1489), he declared that since the universe is infinite, it can have neither a center nor a circumference; rather, any point could be viewed as the center, just as to an observer at sea the horizon seems to be equally distant in all directions, regardless of the observer's position. In this Cusa was no doubt influenced by his fascination with mathematical infinity, with the infinity of number and of endless division. However, Cusa's universe, like that of Giordano Bruno who followed him, was based not so much on scientific reasoning as on theological speculation: the universe is infinite because the omnipotence of the Creator could not tolerate bounds. Thus Cusa, while denying the earth its central position in the universe, cannot be regarded as a true precursor of the Copernican revolution, even though his influence on future theological thought was considerable.

Vague hints at an infinite universe came from other quarters. Georg von Peurbach (1423–1461), a German astronomer and mathematician, based his cosmology on the Ptolemaic model, but to the outermost sphere of the fixed stars, the firmament, he added another sphere, the *primum mobile,* which moved all the others. And Petrus Apianus (1495–1552), a contemporary of Copernicus, added yet another sphere, the *Empyrium,* the abode of God (Fig. 24.1). Again these "models" of the universe were almost entirely theological creations, meant to embellish and glorify the Aristotelian system. If doubts were ever raised about the validity of such a world picture, they were carefully kept behind closed doors; no one dared to challenge in public the official teaching of the Roman Catholic church, which dogmatically adhered to the old system. In some cases, models were proposed which were a regression even from the old system; thus Peurbach actually believed

Figure 24.1. *The universe according to Petrus Apianus's* Cosmographia *(1539). The surrounding sphere is the* Empyrium, *the abode of God. From:* Cosmographia *by P. Apianus. Reprinted from* Man's View of the Universe *by Gerald Tauber. Copyright © 1979 by Crown Publishers, Inc. Used by permission of Crown Publishers, Inc.*

191

Figure 24.2. *The Copernican universe. From* De revolutionibus orbium coelestium 1543).

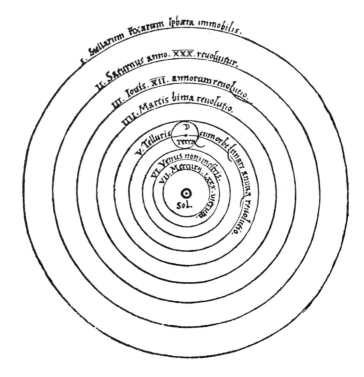

in the existence of crystalline spheres, an idea that even Ptolemy did not quite adopt. And yet, by adding more and more outer spheres to the system, we find at least an indirect admission that the universe may be much larger than had been thought, if not actually infinite.

It is in this light that Copernicus's revolution must be judged. Born in Poland in 1473, Nicolaus Copernicus (or Niklas Koppernigk in the Polish version) first studied astronomy at the University of Cracow, and then moved to Italy, where he rounded up his education in medicine and law. But his main interest was astronomy, and when he returned to his native country in 1507, he took up the position of the canon of the church of Frauenburg, a position which left him ample time for his astronomical studies. Copernicus was to stay in Frauenburg for the rest of his life, and it was there that he wrote his first work, *Commentariolus* ("Comments"). Meant merely as a collection of notes for his friends, the work nevertheless summarized Copernicus's new outlook of the universe. Foremost among his seven propositions are the assertions that

1. the sun, and not the earth, is at the center of the universe;

2. all planets—including the earth—move around the sun;
3. it is the rotation of the earth about its axis—and not the rotation of the firmament around the earth—that causes the alternation of day and night.

These three propositions would become the cornerstone of a new cosmology, the greatest scientific revolution in man's history up until then. Yet equally important, though often neglected, is the fourth of Copernicus's propositions, which says, in his own words: "The heavens are immense in comparison with the earth." As he put it, even the radius of the earth's orbit around the sun is "nothing in comparison with the sphere of the fixed stars."

Here, then, is the essence of the Copernican or heliocentric system (Fig. 24.2). Judged by its own terms, it was not really as revolutionary as is often thought. True, Copernicus demoted the earth from its central position in the universe and replaced it by the sun. But his universe was still a contraption of circles and spheres—the thought that any other shape might play a role in the planetary system had to wait until Kepler—and it was still a finite universe.[1] However, the apparent absence of any measurable parallax of the fixed stars—which hitherto had been attributed to the earth's state of rest—convinced him that his universe must be vastly larger than anyone had thought before; specifically, he concluded that the sphere of the fixed stars should be at least a thousand times farther than the earth's distance from the sun, and at least seventy-five times farther than Saturn's. Thus he was

This ball every 24 hours by naturall, uniforme and wonderfull slie and smooth motion rouleth rounde, making with his Periode our naturall daye, whereby it seems to us that the huge infinite immoveable Globe should sway and tourne about.
□ Thomas Digges (1546–1595)

---

[1] Copernicus's chief goal in proposing his system was to offer a more convenient explanation for the motion of the planets, and especially their occasional retrograde motion (see p. 186), than that given by Ptolemy's system. In the latter, the notion of epicycles plays a central role. An *epicycle* is a circle whose center moves along the circumference of another circle. (The resulting curve looks somewhat like a circular coil; it can be produced with the aid of a spirograph.) Ptolemy used these epicycles to account for the seemingly irregular motion of the outer planets Mars, Jupiter, and Saturn; but since the observational data did not quite fit the system, more and more epicycles were added, making the system extremely complicated. Copernicus showed that by referring all planetary motion to the sun, rather than to the earth, one can dispense with the epicycles. (He retained them, however, to account for the varying velocities of the planets, a phenomenon which was completely explained a century later when Johannes Kepler discovered that the planets move around the sun in ellipses.) In short, Copernicus's system—at least in its original conception—was not much more than a mathematical theory—a far cry from the profound philosophical interpretations that were given to it by later generations. (It will be remembered that Aristarchus, in the third century B.C., had already envisioned a heliocentric system.)

193

well aware of the immense gap which separates the boundary of our solar system from the realm of the fixed stars. The credit for shattering the age-old picture of a small universe, one that can be measured in terms of terrestrial distances, goes thus to Copernicus, and to him alone. If he did not actually admit an infinite universe, we may forgive him: "The man who took the first step, that of arresting the motion of the sphere of the fixed stars," says Alexander Koyré, the noted science historian, "hesitated before taking the second, that of dissolving it in boundless space; it was enough for one man to move the earth and to enlarge the world so as to make it immeasurable—*immensum;* to ask him to make it infinite is obviously asking too much."[2]

For all the aura of a revolutionary zealot that Copernicus's name evokes, the canon of Frauenburg was a quiet and withdrawn man, and nothing was farther from his mind than thoughts of changing the world. He compiled his work in a book, in six volumes, under the title *De Revolutionibus* ("On the Revolutions," the term here meaning "rotations"), which was no doubt modeled after Ptolemy's *Almagest.* Most of this work (the main part was completed in 1533) deals with mundane astronomical matters such as spherical trigonometry, the theory of eclipses, and an update of Ptolemy's star catalog. Only the first volume contains an outline of his new theory (even though much of the subsequent material depends on this theory). But Copernicus was reluctant to publish this monumental work, a summary of his lifetime work, fearing no doubt that it might provoke the Catholic church. It was only with the relentless prodding of his few disciples that he finally yielded, but the publication process started slowly and was interrupted several times. When the work finally went to print, its author was already an old and sick man, barely able to correct the galley proofs, including a preface written by the editor which, in effect, denied everything that Copernicus had said in the book.[3] It was

But there can be no movement of infinity and of an infinite body, and therefore no diurnal revolution of that vastest Primum mobile.
□ Thomas Digges

---

[2] Alexander Koyré, *From the Closed World to the Infinite Universe,* Johns Hopkins University Press, Baltimore, 1974.

[3] The editor, Andreas Osiander (1498–1552), was a Lutheran minister and active in the reform movement. In writing the controversial preface, he evidently wanted to protect himself from any accusations of heresy. (Luther himself was firmly against the new theory.) In any event, the effect was to damage Copernicus's reputation as a scrupulous fighter for the truth, since it was assumed that the preface was his. It was only in 1609 that Kepler discovered in a copy of *De Revolutionibus* a note which identified the true author of the preface. By that time, however, the damage was already done, and it took many years before Copernicus's reputation was restored.

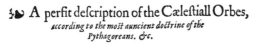

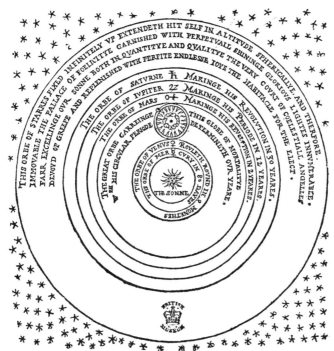

Figure 24.3. *Thomas Digges's universe, probably the first explicit reference to an infinite universe. From* Leonard Digges, Prognostications Euerlasting (*London, 1576*).

a few hours before his death that one of the first copies of *De Revolutionibus* was brought to his bed. He died on May 24, 1543.

Copernicus's new cosmology now had to begin its struggle for acceptance. Initial reception to it was cool, though Copernicus was spared the humiliation that would befall Galileo a century later. One of the few proponents of his system was the English astronomer Thomas Digges (1546–1595), who in 1576 published a book in which he not only adopted Copernicus's system, but also advocated an infinite universe, making him the first professional astronomer to do so. His universe is illustrated in a figure (Fig. 24.3) which clearly shows the sun at the center, surrounded by the orbits of the six planets (the third from the sun being the earth). Outside the outermost orbit, that of Saturn, there is a large gap, beyond which is the realm of the fixed stars. Inside the gap Digges inserted an inscription which declares: "This orbe of starres fixed infinitely up extendeth hit self in altitude sphericallye and therefore immovable . . ." The word "therefore" is inter-

195

Figure 24.4. *The universe of William Gilbert. From his* De mundo nostro sublunari (*Amsterdam, 1651*).

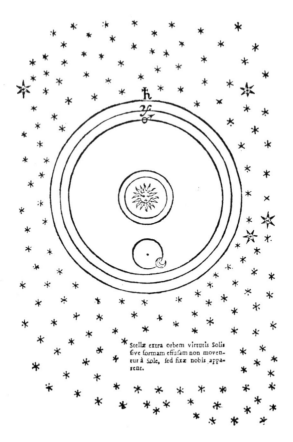

Stellæ extra orbem virtutis Solis sive formam effusam non moventur à Sole, sed fixæ nobis apparent.

esting—it shows that Digges realized the absurdity of letting an infinite universe revolve around a tiny earth, or even a sun. The inscription then pays the obligatory tribute to the greatness of God's creation, describing the stars as "shininge glorious lightes innumerable" and their abode as the "court of coellestial angelles." It is hardly surprising that Digges, the astronomer, had to reassure himself by resorting to the familiar theological themes so popular in his days. We witness here the intellectual struggle through which the human mind had to go before it could accept the idea of a boundless universe.

An even greater confusion of mysticism and rational thought we find in the tragic life and death of Giordano Bruno (1548–1600). Bruno's name will forever be remembered for his martyrdom—a man who made the supreme sacrifice for refusing to compromise his beliefs. It must therefore come to us as a disappointment that Bruno was neither a scientist nor a true philosopher,

196

but more of a preacher, perhaps even a troublemaker, who wandered across Western Europe, preaching his unorthodox views, and bringing upon himself—perhaps intentionally so—the wrath of the Catholic church. Born to poor parents, Bruno entered the Dominican monastery at Naples at the age of fourteen, and there he first read Copernicus's *De Revolutionibus*. He immediately became an ardent believer in the new cosmology—an act which suited his independent, almost rebellious nature. Resenting blind obedience to the Church's doctrines, he found in Copernicus a figure to extol, a symbol of defiance who faced the Church and challenged its ironclad traditions. (As we have seen, nothing was farther from Copernicus's mind.) But whereas Copernicus, the mathematician and astronomer, based his teaching on strict rational arguments, Bruno, the dreamer, took his master's theory and expanded it spiritually to embrace the whole universe. Whereas Copernicus's central doctrine dealt with the motion of the earth, Bruno envisioned an infinite universe, populated by an infinity of stars like our sun, each surrounded by planets on which intelligent beings thrive and prosper. Infinity was Bruno's motto—infinity of space and time, spiritual infinity as well as physical infinity. His speculations—one must almost call them flights of fancy—carried him even beyond our own universe, to the realm of God. But unlike his predecessors, he seldom resorted to traditional Christian symbolism; he was much closer to the Wisdom literature of the Old Testament, and at times he sounded almost like a devout kabbalist. Describing God's eternal wisdom, he likened it to the radiance of infinite light, which "descendeth to us by the emission of rays and is communicated and diffused throughout all things." Time and again he came back to his vision of infinity, of a universe on a grand scale, boundless in space and eternal in time, a universe populated by countless suns like our own: "There is a single general space, a single vast immensity which we may freely call Void: in it are innumerable globes like this one on which we live and grow. This space we declare to be infinite."

Among Bruno's works, the best known is *De l'infinito universo et mondi* ("On the Infinite Universe and Worlds"), first published in Venice in the year 1584. The work is in the form of a dialogue among four "speakers," of whom Philotheo is Bruno in disguise. It opens with Elpino asking his colleagues: "How is it possible that the universe can be infinite?", to which Philotheo retorts with the counterquestion: "How is it possible that the universe can be finite?" And so the dialogue goes on and on, until finally Philotheo convinces the others of his view. Here and elsewhere

It is, I say, impossible that I can with any true meaning assert that there existeth such a surface, boundary or limit, beyond which is neither body, nor empty space, even though God be there.
□ Giordano Bruno (1548–1600), as Philotheo in *De l'infinito universo et mondi*

It is with a burning enthusiasm—that of a prisoner who sees the walls of his jail crumble—that he [Bruno] announces the bursting of the spheres that separated us from the wide open spaces and inexhaustible treasures of the ever-changing, eternal and infinite universe.
□ Alexander Koyré, *From the Closed World to the Infinite Universe*

197

There is a single general space, a single vast immensity which we may freely call Void; in it are innumerable globes like this on which we live and grow. This space we declare to be infinite; since neither reason, convenience, possibility, sense-perception nor nature assign to it a limit. In it are an infinity of worlds of the same kind as our own. For there is no reason nor defect of nature's gifts, either of active or of passive power, to hinder the existence of other worlds throughout space, which is identical in natural character with our own space.
□ Giordano Bruno, *De l'infinito universo et mondi*

in Bruno's works we find a strange mixture of spiritualism and sound reasoning. Thus, justifying the notion that an infinite universe can have neither a center nor a circumference, he says: "To a body of infinite size there can be ascribed neither centre nor boundary . . . Just as we regard ourselves as at the centre of that equidistant circle, which is the great horizon and the limit of our own encircling ethereal region, so doubtless the inhabitants of the moon believe themselves at the centre [of a great horizon] that embraces this earth, the sun and the other stars, and is the boundary of the radii of their own horizon. Thus the earth no more than any other world is at the centre." Here, in a perfectly logical argument, Bruno foresaw the principle of relativity, which would play such a central role in physics three centuries later. But in his poem, *De immenso et innumerabilibus,* we find an argument of quite a different kind: "The One Infinite is perfect; simply and of itself nothing can be greater or better than it. This is the one Whole everywhere, God, universal nature. Naught but the infinite can be a perfect image and reflection thereof, for the finite is imperfect; every sensible world is imperfect, wherefore evil and good, matter and form, light and darkness, sadness and joy unite, and all things are everywhere in change and motion. But all things come in infinity to the order of Unity, Truth and Goodness; whereby it is named *universum.*" And when he comes back to his beloved theme, the plurality of the world, he speaks with the fervor of an evangelist:

> *Thus is the excellence of God magnified and the greatness of his kingdom made manifest; he is glorified not in one, but in countless suns; not in a single earth, but in a thousand, I say, in an infinity of worlds.*

The story of Bruno's last eight years will forever be among the most moving in the annals of history. Having provoked the Church's authorities in his public preachings, he was chased by the Roman Inquisition and finally fell into their hands. In this he must share the blame, for he willingly accepted an invitation by Giovanni Mocenigo, an agent of the Inquisition, to come to Venice, ostensibly to serve as the latter's teacher. The consequences were inevitable: he was arrested, tried for heresy, and condemned to death. What can sustain a man under this supreme test we shall never know, but perhaps it was his unbounded joy at the grandeur of the universe—*his* universe—that lifted his spirits and made him endure the ordeal. He was burned at the stake in Venice on February 17, 1600, uncompromising in his beliefs to the end.

Nature is an infinite sphere, whose center is everywhere and whose circumference is nowhere.
□ Blaise Pascal (1623–1662)

The Horizons Are Receding　25

Not quite ten years had passed since Bruno's tragic death when an event took place that would completely vindicate him and his master, Copernicus. On January 7, 1610, Galileo Galilei (1564–1642), by then already a renowned scientist, aimed his new telescope at the planet Jupiter. To his amazement, he found the planet surrounded by four small objects, which he correctly identified as satellites circling their parent body. He named them the Medicean Stars, in honor of the Medici family in whose service Galileo hoped to be employed. Here, then, was an entire solar system in miniature—a retinue of small bodies circling a large one—and it gave strong, though indirect, support to the theory of a heliocentric system (which even at that time was far from being universally accepted). Even stronger evidence came when Galileo discovered that Venus, a planet closer to the sun than the earth, exhibits phases like the moon—a convincing proof that it must be circling the sun and not the earth.[1] He then directed his telescope (he called it a "spyglass") at the moon and saw sights never seen before by the human eye—a heavenly body crisscrossed by valleys and mountains, by flat plains, and by "seas"—in short, an imperfect world not unlike our own and a far cry from the perfect crystal spheres of the Greeks. To crown it all, Galileo now turned his

---

[1] This in itself, though, is not a sufficient proof that the earth, too, moves around the sun, as required by the Copernican system. In fact, Tycho Brahe's system, in which the planets move around the sun and the sun in turn moves around the earth, could equally well account for the phases of the inner planets Mercury and Venus.

attention to the Milky Way—that band of diffuse light that can be seen in the sky on a clear, dark night—and found it to consist of innumerable stars not visible to the naked eye. Yet the fact that his telescope could not magnify any of these stars, not even the brightest ones, told him that their distance must be enormous—the first direct evidence that the universe is much larger than anyone had thought before. We can easily imagine his excitement at the vast new spaces that his instrument opened up before him; referring to Aristotle, he wrote in his *Dialogue on the Two Chief World Systems:* "We can discern many things in the heavens that he could not see and therefore we can treat the heavens and the sun more confidently than Aristotle could." With these simple words, Galileo ushered in a new era in astronomy—the era of the telescope.

The result of all these discoveries was a little popular book, *The Starry Messenger* (1610), in which Galileo described in glowing terms his new findings. The book was an instant success and contributed greatly to popularize the science of astronomy. But among his learned colleagues the reaction was less enthusiastic: they saw in Galileo's discoveries a direct threat to their authority, and now set out to silence him. After a long campaign of manipulations and intrigues, they managed to bring the matter before the Church. Galileo, by then already an old man, was summoned to Rome to appear before the Holy Chair. After several hearings he was forced to admit his "guilt" in teaching the Copernican system, and in the end he recanted, no doubt having Bruno's fate on his mind. Legend has it that he exclaimed in defiance, "and yet it does move." He did escape Bruno's fate but only barely so, spending his last years virtually under house arrest.[2] He died on January 8, 1642, embittered by his long ordeal and completely blind.

We must mention briefly the views of two of Galileo's contemporaries on the nature of the universe. Tycho Brahe (1546–1601), the great Danish observational astronomer, devised a planetary system which was a compromise between the Ptolemaic and the Copernican: the five planets move in circles around the sun, which in turn moves around the stationary earth. Tycho used the absence of any observable parallax of the fixed stars as evidence that the earth is immovable; like the Greeks, the possibility of a huge

---

[2] It was during this last period of his life that Galileo wrote a second "dialogue," *Dialogue Concerning Two New Sciences,* in which he expressed some of his thoughts about mathematical infinity and its paradoxes (see Part I).

universe—which would equally well account for this fact—did not suit him. For a while his system was quite popular, all the more so since it could explain the motion of the planets without resorting to a heliocentric system. (In fact, the two systems are mathematically equivalent, as is clear from the relative nature of the motion of one body with respect to another.) But Tycho's main contribution to astronomy was his observations of stellar and planetary positions, which he performed with meticulous precision at his observatory on the island of Hven in Denmark. It was on the basis of these observations that Johannes Kepler (1571–1630) was able to derive his three celebrated laws of planetary motion.[3] These he arrived at after years of struggle, not made any easier by the fact that he was an ardent Pythagorean who was led (or misled) by mystical considerations as much as by sound scientific arguments. He tried to base his planetary laws on the geometry of the five regular solids, which he believed to correspond to the five planets; when that failed, he turned to the laws of musical harmony, assigning to each planet a melody according to its distance from the sun. All the same, Kepler's laws were an achievement of the greatest importance, giving astronomers for the first time a precise, quantitative theory on which to base their calculations. What is more, by discovering that the planets move around the sun in ellipses (Kepler's First Law), he once and for all put to rest the good old crystalline spheres, in effect bringing Greek astronomy to a close. Yet Kepler, like Tycho, could not bring himself to accept the possibility of an infinite universe, and Bruno's fantasies about the plurality of worlds filled him with horror. "Surely," he wrote in his *De stella nova*, "rambling across that infinity can do good to no one."

The time was now ripe for someone to unify the new discoveries into a single, all-embracing theory. This task befell Isaac Newton (1642–1727), who was born on Christmas Day (by the Julian calendar) in the year of Galileo's death. If we devote here only a minor space to Newton, it is only because his life and work have been described so often and so extensively elsewhere. Newton's unique achievement lay in his ability to discover through

---

[3] Kepler's laws are:

1. The planets move around the sun in ellipses, with the sun at one focus of each ellipse.
2. The radius vector joining each planet to the sun sweeps equal areas in equal times.
3. The square of the period of revolution of each planet around the sun is proportional to the cube of its mean distance from the sun.

many seemingly unrelated phenomena a single unifying principle—and to formulate it mathematically. In this sense he is the founder of theoretical physics. He had the insight to realize that a falling apple and a planet moving around the sun are subject to one and the same law—the universal law of gravitation.[4] (He first derived it for the earth–moon system, and then generalized it for any two bodies.) With the enunciation of this law in the *Principia* (1687), modern astronomy came of age.

Together with Newton's three laws of motion,[5] the universal law of gravitation forms the basis of celestial mechanics, and even its replacement, early in our century, by Einstein's general theory of relativity has not diminished its role in matters such as determining the orbits of satellites and spacecraft. But its significance went far beyond celestial mechanics, for if one and the same law applies to all gravitational phenomena, whether here on earth or in the remotest corners of the universe, then one can hardly avoid the conclusion that this universe must be infinite. Newton himself believed not only that the universe is infinite, but that it is homogeneous and isotropic, i.e., of the same construction everywhere and in all directions.[6] Only such a universe, he claimed, can main-

---

[4] The law says: The force of gravitational attraction between any two bodies is directly proportional to the product of their masses, and inversely proportional to the square of the distance between them. (Hence the frequent reference to the "inverse square law.") In mathematical terms,

$$F = \frac{G\, m_1 m_2}{r^2}$$

The value of the proportionality constant G was found in 1798 by the English chemist and physicist Henry Cavendish (1731–1810) in a classical experiment in which he carefully measured the extremely small force of attraction between two heavy masses. The value of G (6.67×10^{-8} in the cm-gm-sec system of units) is one of the fundamental constants of physics.

[5] These laws are:

1. A body at rest remains at rest, and a body in motion remains in motion in a straight line and at constant speed unless an outside force acts on it. (The law of inertia.)
2. When a force acts on a body, it causes the body to accelerate: the acceleration is proportional to the force and inversely proportional to the mass of the body. ($F = ma$)
3. For every action there is an equal and opposite reaction.

[6] Interpreted narrowly, the word "universe" could mean just the material universe, which is probably what Newton had in mind (see his quotation accompanying this chapter). However, Newton also assumed that *space*, taken as a geometric entity, is infinite, isotropic, and homogeneous, i.e.,

tain itself in gravitational equilibrium and avoid collapsing towards its center.

The *Principia* has had an enormous influence on subsequent generations of scientists. With its pedantic exposition of the principles of mechanics, it carried the same authoritative weight in physics as Euclid's *Elements* had in mathematics, and its premises would not be challenged until the beginning of our own century.[7] The success of Newton's theory in explaining all the known gravitational phenomena—the motion of the planets, the phenomenon of tides, and the precession of the equinoxes[8] —convinced most physicists that eventually *all* natural phenomena would be predicted from a small number of fundamental principles and mathematical equations. This "deterministic" view of nature would dominate physics for the next two hundred years until it, too, would be challenged in our century and replaced by a more "probabilistic" view. As for now, it seemed that man was on the verge of penetrating the remotest quarters of the universe and about to unravel its most hidden secrets.

Here truths sublime, and sacred science charm,
Creative arts new faculties supply,
Mechanic powers give more than giant's arm,
And piercing optics more than eagle's eye:
Eyes that explore creation's wonderous laws,
And teach us to adore the great Designing Cause.
□ James Beattie (1735–1803)

---

ordinary three-dimensional Euclidean space. It is this latter assumption that was challenged in 1916 by Einstein's general theory of relativity.

[7] The *Principia,* or in its full title, *Philosophiae Naturalis Principia Mathematica* ("Mathematical Principles of Natural Philosophy"), follows the same rigorous style that characterizes the *Elements,* with definitions, axioms, and propositions all listed in a careful, logical order; no doubt Newton, who had a deep respect for Greek science, was influenced by the tradition set by his ancient predecessors. Surprisingly, the work never uses the calculus, the branch of mathematics which Newton himself had created twenty years earlier and which would soon become an indispensable tool of physics. Instead, the *Principia* relies heavily on geometric arguments, again showing Euclid's influence. The work was first published in Latin in 1687; the first English translation did not appear until 1729, two years after Newton's death.

[8] This phenomenon—a slow rotation of the earth's axis once in 26,000 years—had already been known in antiquity, but had never been adequately explained until Newton. He correctly attributed it to the perturbation of the sun's gravitational pull on the earth caused by the latter not being a perfect sphere. The phenomenon can be compared to a spinning top whose motion is perturbed by an outside force, resulting in a slow precession of its axis.

203

26 A Paradox and Its Aftermath

> . . . I have heard urged that if the number of Fixt Stars were more
> than finite, the whole superficies of their apparent Sphere would be
> luminous. . .
>
> — Edmond Halley (1656–1742)

The scientific community was not the only one to rejoice in the
expanding horizons. Philosophers and authors, naturalists and poets—their imagination fired by the new vistas that the telescope
has opened before them—now set to work to describe the new
cosmology. Indeed, their imagination carried them to realms which
even the most powerful telescope could not reach. Echoing Bruno,
the English philosopher and poet Henry More (1614–1687) wrote
his version of the plurality of worlds:

> The Centre of each severall world's a Sunne. . .
> About whose radiant crown the Planets runne,
> Like reeling moths around a candle light;
> These all together, one world I conceit,
> And that even infinite such worlds there be,
> That inexhausted Good that God is hight,
> A full sufficient reason is to me.

In his poem, *Democritus Platonissans; or, An Essay upon the Infinity
of Worlds Out of Platonick Principles,* published in London in 1646
(apparently long titles were necessary in those days to attract the
reader's attention) More not only rejoices at the thought of an
infinite universe, populated by an infinity of worlds like ours,
but he insists on the eternity of time as well. Having "proved"

> That infinite space and infinite worlds there be:
> This load laid down, I'm freely now dispos'd
> A while to sing of times infinity.

One poet, Edward Young (1683–1765), even foresaw the Big
Bang and the creation of the universe. In *Night Thoughts,* while

reflecting on the immense distances that a beam of light must traverse, he says:

So distant (says the sage) 'twere not absurd
To doubt if beams, sent out at Nature's birth,
Are yet arrived at this so foreign world;
Though nothing half so rapid as their flight.

He, too, sees a connection between infinite space and infinite time:

The boundless space, through which these rovers take
Their restless roam, suggests the sister thought
Of boundless time.

Having been intoxicated by the "mathematic glories of the skies," Young then gives free rein to his imagination, sending it on a journey to the unknown:

Loose me from earth's inclosure, from the sun's
Contracted circle set my heart at large;
Eliminate my spirit, give it range
Through provinces of thought yet unexplor'd . . .
Thy travels dost thou boast o'er foreign realms:
Thou stranger to the world! Thy tour begins.

*   *   *

While the eighteenth century poets were celebrating the "aesthetics of the infinite,"[1] often describing the infinity of the world with an infinity of words, astronomers were rapidly expanding their knowledge of the heavens. In 1781 Sir William Herschel (1738–1822), the German-born Englishman, musician turned astronomer, discovered a new planet, Uranus, circling the sun at twice the distance of Saturn. This was the first enlargement of the solar system since antiquity, and it caused tremendous excitement in the scientific community. More new objects were soon to follow. The first of the asteroids, or minor planets, was discovered in 1801, to be followed by hundreds more in the following years. Dozens of new comets, too faint to be seen with the unaided eye, were being discovered through the telescope, and the number of known satellites of the major planets was steadily increasing

---

[1] I took this phrase from Marjorie Hope Nicolson's book, *Mountain Gloom and Mountain Glory: The Development of the Aesthetics of the Infinite* (W.W. Norton, New York, 1959), in which she shows how the eighteenth century's great discoveries in astronomy influenced the views of philosophers and naturalists, in particular their perception of mountains.

*With what an awful,
world-revolving power,
Were first the unwieldy
planets launched along
The illimitable void! There
to remain
Amidst the flux of many
thousand years,
That oft has swept the
toiling race of men,
And all their labored
monuments, away.*
□ James Thomson (1700–1748)

(Herschel himself found two of Uranus's moons within six years of his discovery of the planet.) Most important of all, astronomers were beginning to turn their attention to the objects beyond the solar system, taking full advantage of the ever increasing power of their telescopes. It was Herschel who conducted the first systematic survey of the sky with his 40 feet long, 48-inch diameter instrument, discovering hundreds of double and triple stars, clusters of stars, and objects of an undefined nature, the so-called "nebulae"—a term indicating their vague, diffuse appearance. Herschel's extensive survey, which was extended by his son John (1792–1871) to the southern hemisphere, signaled a shift in emphasis from the solar system to the fixed stars, a process which would culminate in the 1920s with the discovery of the structure of our galaxy.

Speculations about the Milky Way, of course, go back to antiquity; as we have seen, already Democritus, in the third century B.C., came close to guessing its true nature—a huge conglomerate of faint, distant stars. But to most people the Milky Way was simply what it appears to be to the naked eye—a diffuse band of light, a river of milk or fire. (Indeed the Talmud refers to it as Nahr-di-Nur—a "river of fire"; the modern word "galaxy" comes from the Latin Via Lactea—a "road of milk.") As we recall, it was Galileo who first resolved the Milky Way into innumerable individual stars with his telescope, thereby disproving that it was made of a continuous substance. But the first detailed theory about its structure was published only in 1750 by Thomas Wright of Durham (1711–1786), an English instrument builder and amateur astronomer. In his book, *An Original Theory or New Hypothesis of the Universe,* he suggested that the fixed stars lie between two infinite planes, forming a kind of huge slab of finite thickness that slowly rotates about its center like a vast grindstone (Fig. 26.1). Observing this "slab" from the inside, we naturally see many more stars when looking along its central plane than in any other direction—hence the illusion of a continuous band of faint light that girdles the sky. Wright indeed came very close to the true structure of the Milky Way, except that we now know that it has not only finite thickness but a finite diameter as well, and that there exist innumerable other "milky ways" in the universe.

Yet there remained one persisting question that so far had defied all answers: Just how far away are the fixed stars? Once the idea of crystalline spheres had been abandoned, it became obvious that the stars must be at enormous distances from us, but these

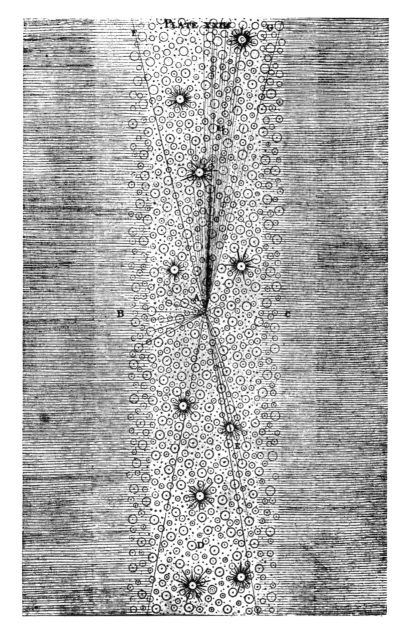

Figure 26.1. *Thomas Wright's picture of the Milky Way, from his* An Original Theory or New Hypothesis of the Universe (*1750*).

distances were still estimated in terms of solar-system dimensions, and in any case no one could offer much more than a wild guess. It was the German astronomer Friedrich Wilhelm Bessel (1784–1846) who first gave a definitive answer to this pressing question.

Figure 26.2. *The phenomenon of parallax: the motion of the earth* E *around the sun* S *produces an apparent shift in the position of a star* P. *The angle of parallax is* α.

In 1838 he was able to measure the parallax of the faint star 61 Cygni (in the constellation Cygnus, the swan) caused by the earth's motion around the sun (Fig. 26.2). This star was known to have a relatively fast proper motion, indicating that it may be rather close. (A *proper* motion is the actual motion of an object relative to the distant sky, as opposed to an *apparent* motion, which is merely due to the observer's own motion.) Even though the parallax that Bessel measured was extremely small—about half a second of arc (by comparison, the moon's apparent diameter is about half a *degree* of arc, or 1,800 arc-seconds)—it enabled him to calculate the star's distance, using simple trigonometry. His result came as a total surprise: eleven light-years, or more than one hundred million million kilometers. Soon after Bessel's discovery an even closer neighbor was found—the star Alpha Centauri, which, at a distance of 4.3 light-years, is the closest known star to the solar system.[2] Thus for the first time the dimensions of the universe beyond the solar system became known.

Twelve years before Bessel's parallax determination of 61 Cygni, an article appeared in the scientific literature which received little attention at the time despite the remarkable question it posed: Why is the sky dark at night? The author of the article, Heinrich Olbers (1758–1840), is not among the giant names in the history of astronomy, even though his contributions were considerable: the discovery of two new asteroids, Pallas and Vesta, and a comet that bears his name, as well as the rediscovery of the asteroid Ceres after it had been "lost" for several years. But Olbers's name is mainly remembered today for his paper, published in 1826 under the title "On the Transparency of Space," in which he posed his famous paradox. If, as was then generally believed, the universe is infinite in extent, and if we assume that on the average the stars are uniformly distributed throughout space (i.e., that the universe is homogeneous and isotropic, as Newton had stipulated), then the line of sight from the earth in any direction should eventually reach some star. Now the intensity of light we receive from any source follows a simple "inverse square law"—it decreases as the square of the distance between the source of light and the observer.[3] On the other hand, the assumptions just

---

[2] Actually Alpha Centauri is a triple star system, of which the faintest component, Proxima Centauri (discovered only in 1915), is at present only 4.2 light-years away.

[3] We recall Newton's universal law of gravitation, which is also an inverse square law. Despite the similarity, however, the two laws are unrelated: the light-intensity law follows from simple geometric considerations (the

mentioned mean that we can think of the universe as consisting of an infinite number of spherical shells of equal thickness but ever increasing radii, all centered around the observer. (These shells, of course, have nothing to do with the crystalline spheres of the Greeks; they are merely a conceptual device used to facilitate the mathematical analysis of the problem, similar to the dissection of a solid into thin "slices" from which its volume can be calculated.) From elementary geometry it follows that the volume of each shell—and hence the number of stars it contains—*increases* as the square of the radius of the shell. Thus, Olbers argued, the decrease in the intensity of light we receive from every individual star is exactly canceled by the fact that there are more stars at greater distances from us than at shorter distances. Therefore, the total light we receive from each individual shell should be constant, i.e., independent of the radius of the shell. Adding the light we receive from *all* the shells, the result will be infinite. In other words, the night sky, far from being dark, should be ablaze with the light of all the stars in the universe!

To resolve the paradox, Olbers offered a plausible explanation. He suggested that the space between stars is filled with dark clouds of cosmic dust which partially absorb the light we receive from distant stars. This assumption turned out to be correct; but we also know today that the amount of interstellar dust is not quite sufficient to diminish the light from distant stars by the required amount. The correct solution to Olbers's paradox could not have been anticipated in his time, for it involves the validity of the very same assumptions on which the paradox is based: an infinite, homogeneous, and isotropic universe (to which Olbers added another tacit assumption—that the universe is infinitely old, and that it has not changed much since the beginning of time; i.e., a static universe). We now know that the universe is finite and expanding, and it is this expansion that reduces the amount of light we receive from distant stars (actually galaxies) through the phenomenon of "red shift," to which we will turn in the next chapter.

The reaction to Olbers's paper was—silence. Astronomers were too busy digesting the enormous harvest of new observational discoveries to pay much attention to a theoretical problem which seemed, at best, to be no more than a thought exercise. Olbers's paradox was soon forgotten, to be rediscovered only in our own

Beyond these are other suns, giving light and life to other systems, not a thousand, or two thousand merely, but multiplied without end, and ranged all around us, at immense distances from each other, attended by ten thousand times ten thousand worlds, all in rapid motion; yet calm, regular and harmonious— all space seems to be illuminated, and every particle of light a world . . . And yet all this vast assemblage of suns and worlds may bear no greater proportion to what lies beyond the utmost boundaries of human vision, than a drop of water to the ocean.
□ From a nineteenth-century astronomy textbook (*The Geography of the Heavens,* by Elijah H. Burrit, 1845)

---

fact that area increases as the square of the linear dimension), while the law of gravitation is a purely physical law whose reasons are much more subtle.

century, when its full implications would be realized.[4] Meanwhile, the flux of discoveries continued unabatedly. In 1814 the German optician Joseph Fraunhofer (1787–1826) found that when sunlight is passed through a narrow slit and then through a prism, the resulting spectrum is crossed by hundreds of thin dark lines, each forming an image of the slit. This marked the birth of spectroscopy, which was later to become our main tool in exploring the physical structure of the stars—the essence of astrophysics.[5] And in 1846 another major planet was found—this time not by accident, but by a triumph of mathematical ingenuity.[6] With such a wealth of discoveries to keep their minds busy, it is no wonder that astronomers chose to ignore Olbers's paradox. The night sky could remain dark for another century before anyone would be bothered again.

---

[4] It turns out that Olbers was not the first to notice the paradox. Edmond Halley preceded him by a century (see Halley's quotation at the beginning of this chapter), and Newton, in arguing for an infinite, homogeneous, and isotropic universe, was aware of the gravitational analogy of the optical paradox. For a full history of the paradox, see the book *The Paradox of Olbers' Paradox* by Stanley L. Jaki, Herder, New York, 1969. See also the article "Why the Sky is Dark at Night" by Edward R. Harrison, *Physics Today*, February 1974, pp. 30–36.

[5] Some spectral lines had been noticed already before Fraunhofer, but he found more than 500 of them, not only in the solar spectrum but also in that of stars. It remained, however, for Gustav Robert Kirchhoff (1824–1887) to correctly interpret the significance of the Fraunhofer lines. When a chemical element is heated to a sufficiently high temperature, it begins to emit light of a specific color content (spectrum), with bright emission lines at various wavelengths. These emission lines are characteristic of each element and are not altered even if the element is present with other elements in a chemical compound. Thus by analyzing the spectrum of an incandescent object (such as a star), one can deduce the chemical composition of the object. Kirchhoff found that the *dark* lines that cross the solar spectrum are at precisely the same wavelengths as the bright emission lines produced by the various elements. He correctly concluded that these dark lines are caused by the absorption of the corresponding wavelengths by the cooler gases present in the sun's atmosphere. This has been, and still is, our basic source of information about the physical structure of the celestial objects.

[6] It had been known for some time that there were discrepancies between the calculated orbit of Uranus and the orbit actually observed. When all attempts to reconcile the two orbits had failed, it began to appear as though an unknown, trans-Uranian planet were perturbing Uranus from its "correct" orbit. Work was begun independently in England by John Couch Adams (1819–1892) and in France by Urbain Jean Joseph Leverrier (1811–1877) to calculate the position of the unknown planet. The story of the search that followed, with its ultimate success—the planet was found on September 23, 1846, by Johann Gottfried Galle (1812–1910) of Germany almost exactly at the predicted position—has all the

---

elements of a human drama, with the main players, Adams and Leverrier, rather than indulging in a bitter dispute over priority (as Newton and Leibniz had over the discovery of the calculus), acknowledging each other's contribution and, in the end, becoming befriended. Quite apart from the human side, the discovery of Neptune showed once again the power of mathematics not only to explain known phenomena but also to predict unknown ones. A fascinating account of the event can be found in the book *The Discovery of Neptune* by Morton Grosser, Dover Publications, New York, 1979.

211

27 The Expanding Universe

It is fairly certain that our space is finite though unbound. Infinite space is simply a scandal to human thought.

— Bishop Barnes

If you go out on a clear, moonless autumn night and look up to the constellation Andromeda, your eyes may catch a glimpse of a faint, diffuse smudge of light. In appearance it cannot compete with some of the more spectacular sights of the sky, such as the surface of the moon or the rings of Saturn, and even a telescope will not reveal much detail: you can make out the outline of an elliptical structure with a central condensed core, but that is about all. Yet before you dismiss this object as insignificant, stop and think: you are looking at the Great Nebula in Andromeda, a sister galaxy to our own Milky Way which, at a distance of 2,000,000 light-years, is the farthest object the unaided human eye can see. To all purpose and extent, you are looking at infinity.[1]

It was in the year 1924 that the Great Nebula in Andromeda was recognized for what it is—an "island universe," a milky way in its own right. In that year, the American astronomer Edwin P. Hubble (1889–1953), using the powerful 100-inch telescope of the Mt. Wilson observatory in Pasadena, California (then the largest in the world), was able to identify individual stars in the arms of that nebula. Thus a mystery that had puzzled astronomers throughout the nineteenth century was finally solved.

In the beginning God created the heaven and earth.
□ Genesis 1:1

Ever since Herschel's survey of the sky in the early years of the nineteenth century, the nature of the nebulae—those diffuse, fuzzy patches of light that seem to dot the sky at random—had

---

[1] Even the largest telescopes don't reveal much detail of extragalactic objects, unless one uses photography. The spectacular view of the Andromeda Galaxy shown in Fig. 27.1 is the result of a long time exposure.

212

Figure 27.1. *The Great Galaxy in Andromeda. A sister galaxy of our own Milky Way, it is the farthest object that the naked eye can see. Its distance from us is about two million light years. A Palomar Observatory Photograph.*

been an enigma to astronomers. Some nebulae, such as that in the constellation Orion, were clearly vast clouds of gas, illuminated by some bright nearby stars. Others were themselves clusters of stars, often having a globular structure and containing up to several thousands of stars. (The magnificent globular cluster M13 in Hercules is the best known of these.) For a while these nebulae were regarded as a nuisance to astronomers, since their hazy appearance could easily be mistaken for a comet. (At the time, the search for new solar system objects was still regarded as the main task of astronomy.) Indeed, to avoid such a confusion, the French astronomer Charles Messier (1730–1817) between the years 1781 and 1784 compiled a catalog of 103 nebular objects, each with its exact position and a description of its appearance in the tele-

213

scope. Messier's catalog is still in use today—the Andromeda nebula, for example, is listed as M31—even though the number of known nebulae has since increased enormously: some 15,000 were known by the year 1900.

Of the objects in Messier's catalog, about one-third showed a clear elliptical shape when viewed through the telescope, making them distinct from the fuzzy, amorphous gaseous nebulae. Some even showed faint spiral arms extending from their elliptical cores, like the wake of a rotating water sprinkler. Yet not even the most powerful telescopes of the nineteenth century were able to resolve these elliptical nebulae into individual stars—a fact that gave rise to one of the greatest controversies in modern astronomy. Most astronomers believed that these nebulae were part of our Milky Way, located perhaps at its outermost fringes. Some, however, speculated that they might be remote milky ways similar to our own and at enormous distances from us. Notable among the proponents of this view was the philosopher Immanuel Kant (1724—1804), who in 1755—just five years after Thomas Wright published his theory of the Milky Way—made the bold suggestion that the entire universe consists of numerous "island universes" or galaxies, as they would later be called.[2] This, of course, was reminiscent of Giordano Bruno's idea of an infinite universe, populated by an infinite number of stars like our sun—except that Kant's vision was on a vastly grander scale. Like Bruno, Kant did not support his views with hard observational evidence, but history would prove him right.

Nothing could be done to resolve the mystery of the elliptic nebulae until more powerful telescopes would be able to reveal their nature, and this had to wait until our own century. Hubble's discovery in 1924 of the existence of individual stars in the arms of the Andromeda galaxy strongly tipped the balance in favor of Kant's theory, but the final and irrefutable proof that the elliptic nebulae were extragalactic objects came when, in that same year, he was able to identify a special type of stars in that nebula from which its distance could be inferred. This was the class of stars known as the *Cepheid variables,* so named after the constellation Cepheus, in which the first of these stars was found. A variable star, as the name suggests, is a star whose brightness varies over time. Some variable stars change their brightness in an irregular, unpredictable way, while others are periodic variables, their light curve repeating itself regularly like a reliable clock. The reason

---

[2] The phrase "island universes" was coined by the German naturalist Baron Alexander von Humboldt (1769–1859).

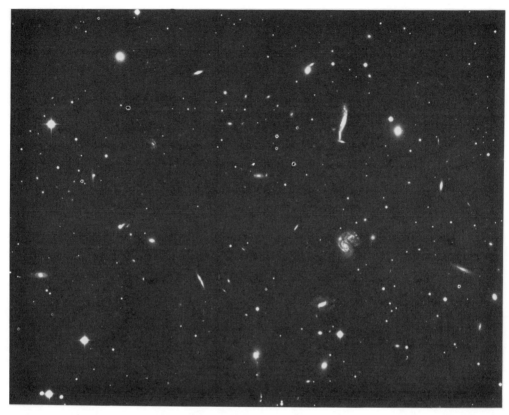

Figure 27.2. A cluster of galaxies in the constellation Hercules. Except for the sharply defined images (representing stars in our own Milky Way), every object in this photograph is a galaxy. All the galaxies in the cluster are held together by the force of gravitation. The cluster (not to be confused with the globular star cluster M13, also in Hercules but lying within our Milky Way) is about seven hundred million light years away. A Palomar Observatory Photograph.

for such a regularity may simply be that a darker star periodically passes in front of its brighter companion, partially eclipsing it from our view. More often, however, the variation is due to some physical process that takes place within the star itself—such as a periodic pulsation of the star's diameter, as if it were breathing. The Cepheid variables belong to this last type. What makes them unique, however, is the fact that their period is directly related to their brightness: all Cepheids of the same period have approximately the same absolute brightness. Thus, by observing the period of a Cepheid variable, we can infer its absolute brightness—its actual output of light (also known as its luminosity). Then, by comparing the absolute brightness to the *apparent* brightness— the brightness we actually observe—we can deduce the distance of the star, using the inverse square law. Thus the Cepheid variables serve as a kind of cosmic yardstick, with the aid of which astronomers were able to determine the distances of numerous celestial objects. When Hubble in 1924 found several Cepheid

215

variables in the arms of the Andromeda nebula, he was immediately able to determine its distance—about one million light-years, putting it well outside our Milky Way.[3] This estimation has since been revised and doubled, but it was Hubble's original discovery that firmly established the existence of extragalactic objects in the universe.

Hubble's determination of the distance of the Andromeda nebula was a breakthrough comparable in its significance to Bessel's parallax determination of the distance of Cygni 61 a century earlier.[4] Soon, however, there followed a discovery which would dwarf even this achievement, and it was again Hubble who made it. When in 1929 he studied the spectra of several extragalactic nebulae (from now on we will simply call them galaxies), he found that their spectral lines were all shifted relative to their natural positions in the spectrum. Moreover, for the great majority of the galaxies, the shift was towards *longer* wavelengths—that is, towards the red end of the spectrum. The meaning of this discovery was immediately clear to Hubble: the galaxies are receding from us.

The universe is not bounded in any direction. If it were, it would necessarily have a limit somewhere. But clearly a thing cannot have a limit unless there is something outside to limit it . . . In all dimensions alike, on this side or that, upward or downward through the universe, there is no end.
□ Lucretius (95–55 B.C.)

To understand the reason for this conclusion, we must digress for a moment to the year 1842, when the Austrian physicist Christian Doppler (1803–1853) discovered the famous effect that bears his name: when a wave-transmitting source is moving towards or away from an observer, the waves received by the observer are shorter or longer, respectively, than those transmitted by the source. A common demonstration of the Doppler effect is afforded when an ambulance with its sirens turned on is approaching and then passing us: we hear a sudden lowering of the pitch, caused by an increase in the wavelength of the sound waves. Doppler's original discovery referred only to sound waves (in fact, he tested it with an orchestra seated on a moving railroad car), but it was soon realized that an analogous effect should apply to any wave phenomenon, including electromagnetic waves (i.e., light).[5] This was verified in 1868 when Sir William Huggins (1824–1910)

---

[3] It was just a few years earlier that Harlow Shapley (1885–1972) was able to determine the structure of our own galaxy—a dish-like conglomerate of some 100 billion stars, 100,000 light-years across and having a central bulge that marks the galactic core. Our sun is located in one of the arms of the galaxy, some 30,000 light-years from its center—in contradiction to Herschel's belief that it is close to the galactic center.

[4] At such immense distances the method of parallax becomes totally ineffective—in fact, it can be used with any degree of certainty only up to about 100 light-years. In general, stellar distances are liable to a wide margin of error, and it is not uncommon to find uncertainties of 50% or more in their determination.

discovered that the spectral lines of several stars were shifted from their normal positions in the solar spectrum. He correctly interpreted this as a Doppler shift caused by the star's motion towards or away from the earth. Since there is a simple formula relating the change in wavelength to the velocity of the source, the radial velocities (the speeds along the line of sight) of stellar objects could be determined with great accuracy. In fact, the Doppler effect has since been used to determine the speeds of anything from highway traffic to orbiting spacecraft.

Hubble, in his historic discovery of 1929, not only found that the galaxies are receding from us, but also that their speeds of recession are roughly proportional to their distances from us: the more distant a galaxy is, the greater its recession speed.[6] This

I am undecided whether or not the visible Milky Way is but one of countless others all of which form an entire system. Perhaps the light from these infinitely distant galaxies is so faint that we cannot see them.
□ Heinrich Lambert
(1728–1777)

---

[5] Although qualitatively similar, the acoustic and optical Doppler effects differ quantitatively due to the different circumstances under which they take place: sound waves travel in a material medium (the air), while electromagnetic waves are transmitted by the electromagnetic field itself. The formulas for the two effects, which we give here in terms of frequencies rather than wavelengths, are:

$$f = \frac{f'}{1 + v/c} \qquad \text{(the acoustic Doppler effect)}$$

$$f = \frac{f'(1 - v/c)}{\sqrt{1 - (v/c)^2}} \qquad \text{(the optical Doppler effect)}$$

In both formulas, f denotes the frequency of the waves as received by a stationary observer, f' is the frequency of the waves as transmitted by the moving source, v is the speed of the source relative to the observer (taken positive if the source moves away from the observer, negative if it approaches him), and c is the speed of the waves. The difference between the two effects, which becomes noticeable only for high velocities compared to the velocity of the waves, can be seen by taking a source that recedes from the observer at half the speed of the waves ($v/c = \frac{1}{2}$): we get $f = (\frac{2}{3}) f' = 0.67f'$ for the acoustic effect, and $f = (\sqrt{3}/3)f' = 0.58f'$ for the optical effect. For a source *approaching* the observer at the same speed ($v/c = -\frac{1}{2}$), we get, respectively, $f = 2f'$ and $f = 1.73f'$. These and some other cases are summarized in the following table, which gives the ratio f/f' for various values of v/c:

v/c	Acoustical	Optical
1/2	0.67	0.58
1	0.50	0
−1/2	2	1.73
−1	∞	∞

[6] Relatively nearby galaxies may deviate from this law due to their own motion within the family of galaxies we belong to, the Local Group. Thus the Andromeda galaxy is actually *approaching* us at a speed of about 200 km/sec. Hubble's law thus applies only to the universe on a large scale, and not to local motions due to the gravitational interaction of neighboring galaxies.

remarkable conclusion, known as Hubble's law, has since been confirmed for thousands of galaxies (whereas Hubble had only a handful to work with—those whose distances he could determine with any degree of certainty). Upon which we may ask: What makes us—i.e., the Milky Way—so special that the galaxies are all running away from us? Do we exert some kind of a repulsive force on the rest of the universe? To quote Sir Arthur Eddington (1882–1944), the eminent English astronomer: "We wonder why we should be shunned as though our system were a plague spot in the universe."[7]

As far as the telescope can reach we find limitless space, unbounded and filled with myriads of such universes.
□ Jenka Mohr, *Sky and Telescope*, October 1934

Ever since Copernicus removed the earth from its supposed center of the universe, the idea that *any* place in the universe might have a privileged position has been categorically rejected. This denial of a privileged status is one of the cornerstones of the theory of relativity, which asserts that the fundamental laws of physics should be the same in all frames of reference, regardless of where the observer chooses to place himself. In cosmology, particularly, this principle has played an important role. Known as the Cosmological Principle, it says that on a large scale, the universe should look the same to an observer here on earth as to his counterpart in a remote galaxy; in other words, that the universe is homogeneous and isotropic. This principle, of course, cannot be proved (just as no physical law can be proved in the mathematical sense of the word), yet the overall distribution of galaxies in the sky strongly suggests it. Moreover, to assume otherwise would be unacceptable on philosophical grounds as well, and even though philosophical arguments should not by themselves be the sole or even the main line of reasoning in science, they nevertheless have always played an important role as a general framework in which our scientific outlook is set. For these reasons, the Cosmological Principle has become the fundamental working basis of modern cosmology.[8]

---

[7] It was through Eddington's efforts that Einstein's theory of relativity was first brought to the public's attention. He was influential in organizing the famous 1919 eclipse expedition which confirmed Einstein's prediction that light rays should be bent near a massive object such as the sun.

[8] However, the qualifying phrase "on a large scale" that was included in the Cosmological Principle is important. When you buy a bag of sugar at the grocery store, you expect its content to be more or less of an even, smooth texture, but this does not preclude the possibility that here and there some sugar will conglomerate into small lumps. Strict homogeneity, like most mathematical concepts when applied to the physical world, is an idealization. We find that the galaxies, like the sugar grains, tend to conglomerate in small groups or clusters, bound together by the same gravitational forces that hold together the solar system (see Fig. 27.2).

218

It follows as a logical consequence of the Cosmological Principle that not only do *we* see the galaxies recede from us at speeds that are proportional to their distances from us, but that an observer in *any* galaxy would see the other galaxies recede from him according to the same law. In other words, *any* galaxy could consider itself to be the center of the universe and discover what Hubble found in 1929. But this can mean only one thing: that *the universe as a whole is expanding.* The situation has often been compared to a balloon whose surface is dotted with many evenly spaced dots. When the balloon is inflated, each dot will move away from all other dots at a speed proportional to the mutual separation between the dots. Thus every dot could claim with equal right that *it* is the center of the universe ("universe" here being the surface of the balloon), and that all other dots are receding from it according to Hubble's law—in perfect compliance with the Cosmological Principle.

Hubble's law can be expressed mathematically by the formula $v = Hd$, where v is the speed of recession of any galaxy from any other galaxy and d is the distance (in light-years) between them. The proportionality factor H, which on account of the Cosmological Principle must be the same for all galaxies at any particular time, is known as Hubble's constant; its value is presently estimated at 50 km/sec per million light-years. This means that a galaxy one million light-years away is receding from us at a speed of 50 km/sec, a galaxy twice as far recedes at twice that speed, and so on.[9] It must be remembered, however, that the values of cosmological parameters, by their very nature, are liable to a wide margin of error, and Hubble's constant is no exception. In fact, its value has been revised several times in the past few decades, and even today one can find estimates that range from one–half to twice the value given above.

Shortly before Hubble's discovery of the recession of the galaxies, another development was taking place on a more theoretical level. Ever since the Greek notion of crystalline spheres had been laid to rest, it has generally been accepted that the material universe is infinite, and that space itself stretches evenly in all directions (i.e., that it is Euclidean). This assumption has been supported by convincing arguments, both philosophical and physical. In fact, as early as the first century B.C. the Roman philosopher Lucretius (95–55 B.C.) argued that if one were to throw a dart over the edge of the universe, it is inconceivable that anything should stop the dart in its motion; hence the universe can have no boundary,

We find them smaller and fainter, in constantly increasing numbers, and we know that we are reaching into space, farther and farther, until, with the faintest nebulae that can be detected with the greatest telescopes, we arrive at the frontier of the known universe.
□ Edwin Hubble (1889–1953)

A closed system of galaxies requires a closed space. If such a system expands, it requires an expanding space.
□ Sir Arthur Eddington (1882–1944)

---

[9] See, however, footnote 6.

The super-system of the galaxies is dispersing as a puff of smoke disperses. Sometimes I wonder whether there may not be a greater scale of existence of things, in which it is no more than a puff of smoke.
□ Sir Arthur Eddington

and it must be infinite. (Clearly Lucretius did not distinguish between "infiniteness" and "unboundedness," a distinction which would later play such a crucial role in the development of non-Euclidean geometry.) Much later Isaac Newton came to the same conclusion, although his arguments were purely physical (we will come back to this shortly). But in 1917, just one year after he completed his general theory of relativity, Albert Einstein (1879–1955) found that his field equations admitted a solution which implied that the universe is finite, yet unbounded!

As we have seen in Chapter 16, Einstein used non-Euclidean geometry to describe the properties of space in a gravitational field. General relativity regards gravitation as a field property, rather than a force acting at a distance, as it was perceived by Newton.[10] Space itself, or more precisely its geometric properties (its so-called "metric"), depends on the strength of the gravitational field at each point, which in turn is a function of the density of matter at that point. Specifically, space becomes "curved"— i.e., assumes non-Euclidean properties—near a massive object like a star or galaxy. When Einstein applied this principle to the universe as a whole, he found that its average density is sufficient to cause space to curve back upon itself, resulting in a finite universe. However, this universe is unbounded, in the sense that one can travel indefinitely in any direction without ever reaching a boundary. This immediately reminds us of the surface of our spherical balloon, but it must be remembered that "real" space —the medium in which physical events take place—is four-dimensional, consisting of the three spatial dimensions (length–width–height) and time. In fact, the theory of relativity regards space and time as one inseparable entity, the *space–time continuum,* in which every point represents an "event." It is in this context that Einstein's universe must be regarded.

In developing his model of the universe, Einstein encountered a serious difficulty. According to the Newtonian theory of gravitation, only an infinite, homogeneous, and isotropic universe could maintain itself in the kind of gravitational equilibrium necessary to prevent it from being pulled together towards its center. In fact, this was Newton's main argument in support of his belief that the universe is infinite. But Einstein's universe was *finite.* To enable it to maintain its equilibrium, he was forced to add, against his better judgment, an additional term to his equations, the so-

---

[10] Contrary to widespread notion, Einstein's theory of gravitation does not contradict Newton's, but rather improves it. Indeed, for small velocities and masses, the two theories are in perfect agreement.

called "cosmological term." Unlike Newtonian gravitation, which is a force of attraction, the cosmological term represented a *repulsive* force whose effect, however, would be felt only at great distances. At the time there was no observational evidence of any kind to support the existence of such a force, and Einstein included it only with great reluctance. What is more, the additional term marred the aesthetic simplicity of his original equations, and considerations of simplicity always ranked high in his world picture. Later, when Hubble discovered the recession of the galaxies, there was no more need for the cosmological term, and Einstein happily discarded it, calling its original inclusion the "biggest blunder" he had ever made.

The reason for Einstein's "difficulties at infinity" (to quote again Eddington) was that his model of the universe was static; it did not allow for any large-scale evolution of the universe in time. This was, of course, in line with the observational facts as they were known in 1917, twelve years before Hubble's discovery. But shortly after Einstein published his results, two other models, both based on general relativity, were proposed: one by the Dutch astronomer Willem de Sitter (1872–1934), and the other by Alexander Friedmann (1888–1925), a Russian physicist. Both models indicated the possibility that the universe may be expanding, and when this was confirmed by Hubble, they achieved instant fame. If Einstein's universe corresponds to a static balloon, de Sitter's and Friedmann's are represented by a balloon which is continually expanding.

If the universe is expanding, there must have been a time when it was much smaller than it is today—a time when the galaxies were very close to each other. By extrapolating backward in time, it should be possible to estimate the time when the galaxies were so close together that they formed one continuous body—the embryo from which the universe was born. This primordial body contained all the matter and radiation in the universe and was concentrated in an extremely small volume, a "singularity" of infinite density. Then a gigantic explosion occurred which set the universe on its present course of expansion. This event, the Big Bang, was the beginning of everything—of matter as we know it today (i.e., the stable elements), of energy, and of space and time. From the presently accepted value of Hubble's constant it is now believed that this event took place sometime between twelve and eighteen billion years ago.

The details of this scenario have been described many times in recent years, both in the scientific literature and in popular works, and we may leave them aside here. Brief mention must,

There was just one place where [Einstein's] theory did not seem to work properly, and that was— infinity. I think Einstein showed his greatness in the simple and drastic way in which he disposed of difficulties at infinity. He abolished infinity. He slightly altered his equations so as to make space at great distances bend round until it closed up. So that, if in Einstein's space you keep going right on in one direction, you do not get to infinity; you find yourself back at your starting-point again. Since there was no longer any infinity, there could be no difficulties at infinity. Q.E.D.
□ Sir Arthur Eddington

however, be made of a rival theory which achieved prominence in the 1940s. Known as the steady state theory, it claimed that the universe has always existed in essentially the same form as we know it today. To account for the recession of the galaxies, the steady state theory assumed that matter is continually being created—not from other matter or even from energy (according to the theory of relativity, matter and energy are equivalent)— but from *nothing*. Aside from the fact that the rate of creation of matter necessary to keep the mean density of the universe at its present level is so low as to practically rule out any possibility of detecting it (about one hydrogen atom per cubic centimeter every billion years), the notion that matter should be created from nothing contradicts one of the most cherished principles of physics, the conservation of mass and energy. Despite these obvious short-comings, the steady state theory at first gained considerable support, mainly because it avoided a problem which for a while had plagued the Big Bang theory. The originally estimated value of Hubble's constant was much higher than the one now accepted, and this put the initial estimate of the age of the universe at only two billion years. But radioactive dating of rocks had already proved that the age of our own earth must be at least 3.6 billion years. (The presently accepted value is 4.5 billion years.) "It was embarrassing to find the earth twice as old as the universe," wrote the astronomer Gerard de Vacouleurs in 1981. The steady state theory, by postulating that the universe is infinitely old, avoided this embarrassment.

Since then, however, the value of Hubble's constant has undergone several drastic revisions, and the presently accepted value makes it clear that the universe is much older than the earth, enabling the Big Bang theory to regain its credibility. But the most convincing evidence in its favor came in 1965, when Arno A. Penzias and Robert W. Wilson discovered the existence of a faint background radiation that pervades the universe in every direction. Known as the three-degree cosmic microwave radiation,[11] it is regarded as the remnant of the huge fireball that created the universe—the Big Bang. The existence of precisely this radia-

---

[11] The name comes from the fact that this radiation is equivalent to a black-body temperature of three degrees Kelvin above the absolute zero. (A black body is an idealized object that can only absorb, but not reflect, the radiation energy impinging on it. This energy is then converted to heat inside the body. In our case the universe itself is the black body, and in this sense one may say that the temperature of the universe is three degrees Kelvin.)

tion had been predicted by several theoretical studies, and its eventual discovery (by accident, as it happened) is regarded as the strongest evidence yet in support of the Big Bang theory.

This, then, is our present picture of the universe: a finite yet unbounded world, created some twelve to eighteen billion years ago in a vast explosion that sent it on its present course of expansion. Will it continue to expand forever? This depends on how much matter it contains. If the total amount of matter in the universe is greater than a certain critical value, this matter will exert enough gravitational pull on itself to slow down the expansion and eventually bring it to a halt, to be followed by a contraction back to the original singularity. Then another Big Bang will create a new universe, and so on forever — a kind of cosmic fulfillment of the Hindu belief in the reincarnation of the soul. If, on the other hand, the amount of matter is less than the critical value, the universe will go on expanding forever, though at an ever slower rate.[12] The determination of the amount of matter in the universe is presently one of the chief tasks of cosmology, and it may be a while before a definite answer can be given. Until then, the question of the future of the universe remains open.

I will give you a "celestial multiplication table." We start with a star as the unit most familiar to us, a globe comparable to the sun. Then—
A hundred thousand million Stars make one Galaxy;
A hundred thousand million Galaxies make one universe.
□ Sir Arthur Eddington[13]

---

[12] The situation is analogous to a stone thrown upwards at a certain initial velocity. For small velocities the stone will reach a maximum height and then fall back to its starting point. Above a certain critical velocity, however (about 11 km/sec), the stone will have enough kinetic energy to overcome the gravitational pull of the earth, and it will escape to infinity. (Instead of changing the stone's velocity, we might think of changing the mass of the earth: the larger the mass, the greater the escape velocity.) In the case of the universe, the initial velocity was imparted to it by the Big Bang.

[13] All the Eddington quotations in this chapter are taken from his book *The Expanding Universe,* Cambridge University Press, New York, 1933, with permission.

28 The Modern Atomists

Where the telescope ends, the microscope begins.
Who is to say of the two, which has the grander view?

— Victor Hugo (1802–1885), *Les Misérables*

In our story of the infinite we have looked mainly at the infinitely large, perhaps because it has received so much attention since Cantor's pioneering work in the 1880s, perhaps also because there seems to be something about the infinitely large that captures the imagination in a way that the infinitely small cannot. This bias, however, is hardly justified. In the history of mathematics the infinitely small has played a role at least as important as its counterpart on the other extreme of the scale. If nothing else, it lies at the root of the notion of continuity, an idea that goes back all the way to the Greeks, whose philosophers heatedly debated the possibility of endless division. And much later, disguised as the infinitesimal, it would become the cornerstone around which the calculus was developed.[1] In any event, from a purely mathematical point of view the distinction between "large" and "small" is not really as fundamental as it may seem, since we can always use the function $y = 1/x$ (or its two-dimensional equivalent, the transformation of inversion) to change the one into the other.

It is, of course, an entirely different story when we come to the natural sciences. Here the difference between the two infinities is the difference between the microcosmos and the macrocosmos,

---

[1] The controversy aroused by the infinitesimals (see Chapter 2) has been satisfactorily settled only recently with the development of a new branch of mathematics known as non-standard analysis. Developed chiefly by Abraham Robinson (1918–1974), it defines the infinitesimal in a rigorous way based entirely on the properties of the real number system. See the article "Nonstandard Analysis" by Martin Davis and Reuben Hersh, *Scientific American*, June 1972.

224

between the subatomic particle and the entire universe. The search for the atom, that elusive smallest particle of which all matter is made, again brings us back to the Greeks. It was Democritus (*ca.* 470–380 B.C.) who first suggested—though on philosophical grounds—that matter is not infinitely divisible. All matter, he believed, is made up of tiny indestructible particles, the atoms (from the Greek *atomos* = indivisible), which by their numerous combinations form the world around us. His idea lay dormant for more than two millennia, until it was revived by John Dalton (1766–1844), who made it the cornerstone of his theory of chemical combinations.

In our own century the atom was to be replaced first by the nucleus, and then by the subnuclear particles. The first hint that the atom may after all be divisible came around 1900 with the discovery of radioactivity by Henri Becquerel (1852–1908). This was followed by the experiments of Ernest Rutherford (1871–1937) in splitting the atom, and then by Sir James Chadwick's (1891–1974) discovery of the neutron in 1932, proving that not only the atom, but also its nucleus, has an inner structure.

Since then, the search for *the* ultimate particle has assumed the proportions of a gigantic race, with the entire scientific community as the participants. Each year a host of new "fundamental" particles are being discovered and announced with great fanfare, only to be later split into still smaller ones. The list is bewildering, as are some of the names given to these new creations: baryons, leptons, mesons, and quarks, to name but a very few. Yet it is still an open question whether the ultimate elementary particle does indeed exist, or whether we are searching in vain for a mere idealized concept, whose existence is no more real than the existence of the mathematical point. Like a child's toy egg, inside which a smaller egg is hidden, and a smaller one inside that, we must contemplate the possibility that matter may never reveal to us its innermost secret.[2]

But in the macrocosmos, at least, this dilemma is spared us. We find in the heaven an orderly hierarchy of things, in which smaller bodies always move around larger ones: moons orbit their parent planets, planets circle around their suns,[3] and stars slowly

The mind of man has perplexed itself with many hard questions. Is space infinite, and in what sense? Is the material world infinite in extent, and are all places within that extent equally full of matter? Do atoms exist, or is matter infinitely divisible?
□ James Clerk Maxwell (1831–1879)

We will have to abandon the philosophy of Democritus and the concept of elementary particles. We should accept instead the concept of fundamental symmetries.
□ Werner Heisenberg (1901–1976)

---

[2] On the other hand, the existence of an "atom" of energy has been firmly established ever since Max Planck (1858–1947) in the year 1900 postulated that energy must exist in multiples of a fundamental quantity, the *quantum.* Later to be called the photon, it became the basis of quantum theory.

[3] Firm evidence of the existence of planets around stars other than the sun has only recently begun to come in.

Infinities and indivisibles transcend our finite understanding, the former on account of their magnitude, the latter because of their smallness; Imagine what they are when combined.
□ Galileo Galilei (1564–1642) as Salviati in *Dialogues Concerning Two New Sciences*

rotate around the centers of their galaxies. Next we find galaxies circling around larger galaxies, forming local groups or clusters of galaxies, then superclusters of galaxies, and finally the entire universe. But have we really reached the end of things? One can, of course, speculate that other universes may exist, perhaps even clusters of universes and so on up the ladder, like in Cantor's endless hierarchy of infinities. Such speculations, however, bring us to the realm of metaphysics. In science we must confine ourselves to the observable world, and in this sense the universe must be regarded as the last rung in our ladder.

---

[1] Normally Pluto, the ninth planet from the sun, is the outermost planet ▷ of the solar system. However, because of its highly elliptical orbit, for a short period during each of its 248-year revolutions around the sun Pluto's orbit crosses that of Neptune. This was the case during Pioneer 10's flight. Pluto's orbit will remain inside Neptune's until the year 2000.

Which Way from Here?   29

On to infinity!

— NBC Evening News, June 12, 1983

On March 3, 1972, the spacecraft Pioneer 10 was launched from its pad at Cape Canaveral, Florida, bound for the giant planets Jupiter and Saturn. In addition to its plethora of scientific instruments, the craft also carried a small, gold-plated plaque with a unique message engraved on it. The plaque (Fig. 29.1) showed a man and woman against the outline of the spacecraft, a sketch of the solar system with the trajectory of the spacecraft indicating where it had originated from, and the positions of fourteen prominent pulsars in our galaxy from which the location of our solar system could be determined. Also shown was a diagram of the hydrogen atom, whose frequency and wavelength of radiation could be used as a universal clock and yardstick. By comparing the wavelength to the number shown in binary code next to the woman's image, an alien scientist could infer the size of those who sent this cosmic messenger. Similarly, by comparing the observed frequency of each pulsar with the data on the plaque, the time that has elapsed since the launch of the craft could be deduced (based on the fact that a pulsar's frequency decreases steadily at a known rate). Thus *Homo sapiens* sent an identifying message across the cosmic ocean, announcing to any alien civilization our existence in this niche of our galaxy.

After more than twelve years of a flawless flight, during which it sent back the first close-up pictures of Jupiter, Pioneer 10 crossed the orbit of Neptune on June 12, 1983, becoming the first man-made object to leave our solar system.[1] Traveling at the incredible speed of over 30,000 miles per hour, its tiny 8-watt transmitter sent its feeble signals back to earth over a distance of more than

Let your soul stand cool and composed before a million universes.
□ Walt Whitman (1819–1892)

Dark—heaving—boundless, endless, and sublime—The image of Eternity.
□ Lord Byron (1788–1824), *Childe Harold's Pilgrimage*

NASA National Aeronautics and
Space Administration

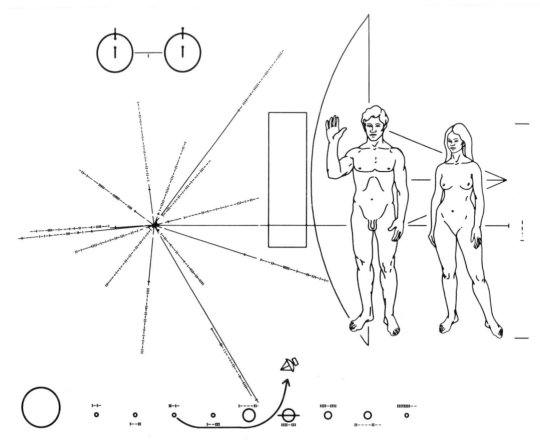

Figure 29.1. *The plaque on board Pioneer 10. Photo courtesy NASA.*

2.8 billion miles, taking them (at the speed of light) 4 hours and 16 minutes to reach us. Pioneer 10 and its twin craft Pioneer 11 are scheduled to send their signals at least until 1994, when their plutonium generators are expected to give out. After that, the two spacecraft will continue their flight through interstellar space, with Pioneer 10 passing close to the star Ross 248 around the year 30,620 A.D. Pioneers 10 and 11 are the closest man has ever come to infinity.[2] And not only infinity in space, but

*Higher still and higher
From the earth thou springest
Like a cloud of fire;
The blue deep thou wingest,
And singing still dost soar,
and soaring ever singest.*
□ Percy Bysshe Shelley
  (1792–1822), *To a Skylark*

---

[2] Two later spacecraft, Voyagers I and II, are also bound to leave the solar system. Each carries a record with the pictures and sounds of our planet, encoded in digital form. For a detailed description of the Voyager message, see *Murmurs of Earth: The Voyager Interstellar Record* by Carl Sagan, Random House, New York, 1978.

228

Figure 29.2 *On to infinity! This picture of a crescent-shaped Earth and Moon— the first of its kind ever taken by a spacecraft—was taken on September 18, 1977, by Voyager I on its journey to the outer solar system. It was taken from a distance of more than 7 million miles from Earth.* Photo courtesy NASA.

also infinity in time—eternity—for the two spacecraft are expected to far outlive our own planet. Long after the earth will have been incinerated by the dying sun when it swells to become a red giant during its final death throes, the two Pioneers, mute and derelict, will sail on unscathed—lonely messengers of a lost civilization.

Since the dawn of history, man has wondered whether he is alone in the universe. The ancients, and later the Roman Catholic church, put our earth at the center of the world, not only physically but morally as well—the one abode of intelligent life in the universe. But this view began to change during the Renaissance, when Nicolaus Copernicus removed the earth from its lofty position and replaced it by the sun. The new cosmology profoundly changed man's view of his place in the universe, and Bruno's vision of an infinite universe, teeming with planets on which intelligent beings are thriving, fired the imagination of subsequent generations. In our time the quest for extraterrestrial life has turned from fanciful speculations to a systematic search, with a fleet of spacecraft roaming through the solar system and visiting new and strange worlds, and giant radio telescopes listening for possible intelligent messages from nearby star systems.

Thus far, these attempts have not born fruit, and the question

They have placed themselves beyond the pale of humanity by transcending the limits God had set for his earthly creatures.
□ Jules Verne (1828–1905), *From the Earth to the Moon*

When we have taken together the sun and all the naked-eye stars and many hundreds of millions of telescopic stars, we have not reached the end of things; we have explored only one island—one oasis in the desert of space. Other islands lie beyond.
□ Sir Arthur Eddington (1882–1944)

229

of whether we are alone in the universe remains perhaps the single greatest mystery yet to be solved. But whatever the answer will turn out to be, the ultimate goal of this grand exploration is to know ourselves and the little planet on which we dwell. Slowly and painfully we are beginning to realize that planet Earth is but a tiny island of habitation in a vast, inhospitable space. For too long have we exploited this island, taking for granted that its resources would last forever. For too long have we polluted its atmosphere and dumped our wastes in its waters. The space age has finally made us aware that this exploitation cannot go on forever, that there is a delicate balance between man and his environment, and that by disturbing this balance we are endangering our very existence on this planet. As Walter Cronkite said so eloquently when the three astronauts of Apollo 8 became the first humans to orbit the moon: "In 1968 we went to the moon and discovered the earth."

. . . Joy and amazement at the beauty and grandeur of this world of which man can just form a faint notion.
□ Albert Einstein (1879–1955); epithet inscribed on his statue at the National Academy of Science, Washington, D.C.

Epilogue

The most beautiful thing we can experience is the mysterious. It is the source of all true art and science.

— Albert Einstein
(1879–1955)

Epilogue

All that we see or seem
Is but a dream within a dream.

— Edgar Allan Poe (1809–1849)

Silently one by one, in the infinite meadows of heaven Blossomed the lovely stars, the forget-me-nots of the angels.
□ Henry Wadsworth Longfellow (1807–1882)

We have come to the end of our journey to the infinite. It has carried us from the "horror infiniti" of the Greeks, through the exultation in an infinite universe during the Renaissance, up to the mathematical breakthroughs of the nineteenth and early twentieth centuries, which finally demystified infinity and put it on a firm basis. We have also followed man's attempts to reach the infinite physically—from the Tower of Babel to Pioneer 10. And we have seen how artists and poets have depicted the infinite, each in his own way. It is this diversity, I believe, that makes the infinite—or any intellectual venture for that matter—so stimulating. Each of us is entitled to our own infinity.

Man's unhappiness, as I construe comes from his greatness; it is because there is an Infinite in him, which with all his cunning he cannot quite bury under the Finite.
□ Thomas Carlyle (1795–1881)

In the final analysis, infinity is a vision, an image. Here is how one fourth-grade student imagined the infinite:

Infinity is a number that is impossible to count to. It is as many atoms as there are on planet earth, or in the solar system. Infinity is how many raindrops there are in a down pour. It is how many pages of homework you get in school . . . It is how far it is to reach the end of a circle. Infinity is how far it is to the farthest star. Infinity is how long it is till there is peace in all the world and universe. [1]

I had my own vision of infinity when standing one morning on top of the great Gateway Arch in St. Louis. Gazing into the

---

[1] By Glen Schuster of Altoona, Wisconsin.

distance, I could see the Great Plains stretching endlessly to the horizon, their flatness disturbed not by the smallest hill. I thought of the thousands of pioneers who crossed the Mississippi at this place, embarking on a westward journey whose end they could not foresee, a journey which must have seemed a voyage to infinity. I had another vision of infinity when, in 1967, I stood on the summit of Mount Katherina, the highest peak in the Sinai Peninsula (8652 feet above sea level). We began the climb shortly after midnight and reached the top just before sunrise. As the first sun rays began to illuminate the desolate landscape below, I could see the shadow of the mountain graze the western horizon, as though emerging from infinity. A spectacular sight it was! The effect must be even more dramatic at sunset, when the mountain's shadow lengthens at an ever faster speed, until it dissolves into nothingness.[2]

Yes, infinity is a vision. To some, it may be but a fleeting cloud, a phantom image of a rainbow. But to others, that same rainbow is real—as real as the colors that make it up. To the mathematician, infinity *is* a reality. In fact, mathematics could hardly exist without it, for it is inherent already in the counting numbers, which form the basis of practically all of mathematics. Moreover, while others have merely described the infinite, mathematicians have made some practical use of it—for instance, in the designing of geographical maps. Perhaps the person who came closest to an understanding of the infinite was Georg Cantor, who realized that infinity should be regarded as something complete, as one whole. But many others shared in the long struggle to come to grips with the infinite. Among them we recall Giordano Bruno, who envisioned an infinite universe and paid for it with his life; Carl Friedrich Gauss, who removed the Euclidean plane from the lofty place it had occupied since the Greek era; and Maurits C. Escher, who depicted the infinite in his prints as no other artist has. Like so many of Escher's pictures, where reality and illusion are so masterly interwoven, each of us must choose his own vision of infinity. Let us, therefore, give the last word to the poet:

In Europe, the frontier was a border, in America it was the horizon; one was where you had to stop, the other stretched as far as you could imagine.
□ Peter Davis in "Hometown U.S.A.," *Family Weekly*, April 18, 1982.

We all live under the same sky, but we don't all have the same horizon.
□ Konrad Adenauer (1876–1967)

---

[2] In 1975, two climbers who reached the summit of Mt. Everest shortly before sunset reported seeing the mountain's shadow stretch to a distance of some 200 miles. The effect should be even more dramatic on the moon, where there is no atmosphere to blur the view. One would see the shadow of the slightest obstruction literally vanish to infinity as the last sun rays disappear behind the lunar landscape.

233

The Infinite

by

Giacomo Leopardi (1798–1837)

This lonely knoll was ever dear to me
and this hedgerow that hides from view
so large a part of the remote horizon.
But as I sit and gaze my thought conceive
intermediate spaces lying beyond
and supernatural silences
and profoundest calm, until my heart
almost becomes dismayed. And as I hear
the wind come rustling through these leaves,
I find myself comparing to this voice
that infinite silence: and I recall eternity
and all the ages that are dead
and the living presence and its sounds. And so
in this immensity my thought is drowned:
and in this sea is foundering sweet to me.

—(Translated from the Italian by Jean-Pierre Barricelli)

Appendix

1. Euclid's Proof of the Infinitude of Primes

The proof that there are infinitely many prime numbers is attributed to Euclid, and is to this day regarded as a model of logical clarity. His proof follows the so-called *indirect* method: We temporarily assume that there is only a finite number of primes, say n. Then we show that this assumption leads to a logical contradiction, and hence the assumption must be false.

Let the n primes be $p_1, p_2, p_3, \ldots, p_n$. We now create a number N by multiplying all these primes together and then adding 1 to the product:

$$N = p_1 p_2 p_3 \cdots p_n + 1$$

Now the number N, by its very construction, is greater than each of the primes $p_1, p_2, p_3, \ldots, p_n$. Moreover, by our assumption N cannot be prime, since we have assumed that the set $p_1, p_2, p_3, \ldots, p_n$ constitutes *all* the primes. So N must be composite, and therefore, by the Fundamental Theorem of Arithmetic, factorable into some of the primes in our set. But if we try to divide N by any of these primes, we will always get a remainder 1 (because of the 1 we added to their product). Thus N is *not* divisible by any of the listed primes. This can only mean one of two things: either N itself is a prime, not listed in our original set, or that N must have among its factors some new prime (or primes) not included in our set. In either case we have a contradiction, because we assumed that our set includes *all* the primes. Therefore our assumption is untenable: the set of primes cannot have an end—it is infinite.

To illustrate, suppose that our set of all primes consists of 2, 3, and 5. The $N = 2 \cdot 3 \cdot 5 + 1 = 31$, which is a new prime not included in our set. On the other hand, if our set consisted of the primes 3, 5, and 7, then $N = 3 \cdot 5 \cdot 7 + 1 = 106$, which is the product of the primes 2 and 53. Thus two new primes must be added to our set. This process can now be repeated: from our new set, consisting (in the first example) of the primes 2, 3, 5, and 31, we can create a new N, namely, $N = 2 \cdot 3 \cdot 5 \cdot 31 + 1 = 931 = 7 \cdot 7 \cdot 19$. Thus two more primes, 7 and 19, are generated. By continuing this process we can actually generate more and more primes, creating a mathematical "chain reaction" of sorts.

2. A Proof That $\sqrt{2}$ Is Irrational

There exist at least three proofs that the number $\sqrt{2}$ is irrational. We will give here an algebraic proof based on the Fundamental Theorem of Arithmetic. This, however, is not the original proof used by the Pythagoreans when they first discovered the irrationality of $\sqrt{2}$; in all likelihood, they gave a proof based on geometric arguments.[1]

We will again follow the indirect method. Assume that $\sqrt{2}$ is rational, i.e., that it can be written as the ratio of two integers:

$$\sqrt{2} = \frac{m}{n} \qquad (1)$$

From this equation we get, by squaring, $2 = m^2/n^2$ or

$$m^2 = 2n^2 \qquad (2)$$

Now since m and n are integers, they can be factored uniquely into their prime factors. Thus, let $m = p_1 p_2 \cdots p_r$ and $n = q_1 q_2 \cdots q_s$. Putting this back into Equation (2), we get

$$(p_1 p_2 \cdots p_r)^2 = 2(q_1 q_2 \cdots q_s)^2$$

or

$$p_1 p_1 p_2 p_2 \cdots p_r p_r = 2 q_1 q_1 q_2 q_2 \cdots q_s q_s \qquad (3)$$

Now among the primes p_i and q_i, the prime number 2 *may* occur (it will occur if either m or n is an even number). If it does

---

[1] See, for example, *An Introduction to the History of Mathematics* by Howard Eves, Holt, Rinehart and Winston, Fourth edition, New York, 1976, p. 65.

occur, it must appear an *even* number of times on the left side of Equation (3) (because each prime there appears twice) and an *odd* number of times on the right side (because 2 already appears there). This is true even if 2 does *not* occur among the p_i's or q_i's; in that case 2 will not appear at all on the left side, while it will appear once on the right side. In either case we have a contradiction: since the decomposition into primes is unique, the prime 2 cannot appear an even number of times on one side of the equation and an odd number of times on the other. Thus Equation (3), and therefore Equation (1), cannot hold true: $\sqrt{2}$ cannot be written as the ratio of two integers, and it must therefore be irrational.

The same proof can be used to show that the square root of every prime number is irrational. But to show that the numbers π and e are irrational, much more powerful methods are necessary. The irrationality of π was proved in 1768 by the Swiss mathematician Johann Heinrich Lambert (1728–1777). More than a hundred years later, in 1882, it was proved that π is not only irrational but in fact *transcendental* (see foonote 1, p. 11). The proof was given by the German mathematician Ferdinand Lindemann (1852–1939) and was thirteen pages long. With it the status of this important number was finally established, ending almost four millennia of speculation and search into its nature.

3. The Convergence of the Geometric Series and the Divergence of the Harmonic Series

To show that the geometric series $a + aq + aq^2 + \cdots$ converges for $-1 < q < 1$, we first consider the *finite* geometric series, or progression:

$$S = a + aq + aq^2 + \cdots + aq^{n-1} \qquad (1)$$

(This series has n terms: a is the *initial term*, and q is the *common ratio*.) Multiplying both sides of Equation (1) by q, we get

$$qS = \quad aq + aq^2 + \cdots + aq^{n-1} + aq^n \qquad (2)$$

[Note that we have moved each term in Equation (2) one position to the right relative to Equation (1).] If we now subtract Equation (2) from Equation (1), all terms except the first and the last will cancel:

$$S - qS = a - aq^n \qquad (3)$$

From here we get $S(1 - q) = a(1 - q^n)$, or

$$S = \frac{a(1-q^n)}{1-q} \qquad (4)$$

This formula gives the sum of the first n terms of the geometric progression in terms of a, q, and n.

Note that in Equation (4) the only term which depends on n is the term q^n. If the absolute value of q is less than 1 (i.e., $-1 < q < 1$), this term will get smaller and smaller as n increases; that is $q^n \to 0$ as $n \to \infty$. Thus, as the number of terms in the series increases without bound, its sum, *provided* $-1 < q < 1$, approaches the limit $a/(1-q)$. We say that the infinite geometric series $a + aq + aq^2 + \cdots$ converges to, or has the sum,

$$S = \frac{a}{1-q} \qquad (5)$$

We must again stress that this formula has a meaning only if $-1 < q < 1$; it is meaningless for values of q greater than 1 in absolute value. For example, for $a = 1$ and $q = 2$, the right side of Equation (5) gives $1/(1 - 2) = -1$, whereas the corresponding series $1 + 2 + 4 + 8 + \cdots$ obviously diverges.

To show that the harmonic series diverges, we make use of the *comparison* technique: we compare the given series with another series, the "comparison series," whose status of convergence or divergence we know. If each term of the original series is smaller than the corresponding term of the comparison series, and if the latter converges, so will the given series. If, on the other hand, each term of the original series is *greater* than the corresponding term of the comparison series, and if the latter *diverges*, so will the original series. We will show that the harmonic series belongs to this latter case.

Let the harmonic series be denoted by S:

$$S = 1 + 1/2 + \underbrace{1/3 + 1/4} + \underbrace{1/5 + 1/6 + 1/7 + 1/8} + \cdots \qquad (6)$$

In each of the bracketed groups of terms, let us replace every term by the *last* term of the group; this will produce a new series S':

$$S' = 1 + 1/2 + \underbrace{1/4 + 1/4} + \underbrace{1/8 + 1/8 + 1/8 + 1/8} + \cdots \qquad (7)$$

Now in doing so, we have replaced each term of the original series by an equal or *smaller* term (for example, $1/4 < 1/3$, $1/8 < 1/5$, etc). Thus every partial sum of S' is smaller than the corresponding partial sum of S: $S'_n < S_n$, or alternatively $S_n > S'_n$, where S_n and S'_n are the first n partial sums of S and S', respectively. But the series S' can be written as

$$S' = 1 + 1/2 + 1/2 + 1/2 + \cdots, \qquad (8)$$

since each bracketed group of terms has the sum $1/2$. This last series obviously diverges. Therefore the harmonic series also diverges.

It should be pointed out that nothing in this proof gives any indication of how fast or slow the harmonic series diverges. This is a typical case of an *existence theorem,* a theorem that establishes a mathematical fact (in this case, the divergence of the harmonic series) without giving any clue as to the numerical values of the quantities involved (here the rate of divergence). Another famous existence theorem is the theorem that $\sqrt{2}$ is irrational: the proof gives no clue about the numerical value of this number. Its value must be inferred from elsewhere, namely from the Pythagorean Theorem (see p. 45). In this sense, the convergence proof of the *geometric* series has a big advantage: it actually produces a formula from which the sum can be found. The geometric series is one of the very few infinite series for which such a formula can be easily found, or even found at all (see Abel's quotation on p. 29).

4. Some Properties of Circular Inversion

We will prove two properties of circular inversion mentioned in the text. We assume that the circle of inversion is the unit circle, i.e., the circle with radius 1 and center at O. We denote this circle by c.

PROPERTY 1: *Inversion transforms lines not passing through O into circles that pass through O, and conversely.*

Proof: Let the line be l (Fig. A4.1). We choose two points on l, the point P closest to O and any other point Q. Let the images of P and Q under the inversion be P' and Q', respectively. Thus, $OP' = 1/OP$ and $OQ' = 1/OQ$, or

$$OP \cdot OP' = OQ \cdot OQ' = 1 \qquad (1)$$

from which we get

$$\frac{OP}{OQ} = \frac{OQ'}{OP'} \qquad (2)$$

But this means that the triangles $\triangle OPQ$ and $\triangle OQ'P'$ are *similar* (the angle $\angle POQ$ being common to the two triangles). Since P is the point on l closest to O, the line OP is perpendicular to l; thus $\triangle OPQ$ is a right triangle with its right angle at P. Therefore,

239

$\Delta OQ'P'$ is also a right triangle, with its right angle at Q', and this is true regardless of the position of Q on the line l. We now use a well-known theorem from Euclidean geometry: The locus of all points from which a given segment is seen at a right angle is a circle whose diameter is the given segment. Thus as Q moves along the straight line l, its image Q' describes a circle k with diameter OP'. This circle passes through O.

Note that in Fig. A4.1 the line l passes outside the circle of inversion c. But our proof is perfectly general—it will hold even if l intersects c (as long as it does not pass through O). In this case, of course, the image circle k will lie partially outside c (see Fig. 12.3).

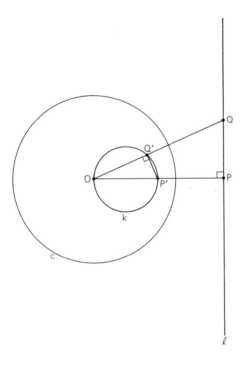

Figure A 4.1

To prove the converse of this property—that inversion carries every circle passing through O into a straight line not through O—we merely have to use the symmetry property of inversion, which says that the words "point" and "image point" can always be interchanged (see p. 90). Therefore, if the point Q' describes a circle through O, its image, which is the point Q, will move along a straight line not through O. This completes the proof.

PROPERTY 2: *Inversion transforms circles not passing through O into circles not passing through O.*

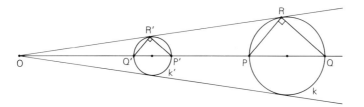

Figure A 4.2

Proof: Let the circle be k (Fig. A4.2), and let us choose three distinct points on k: the point P closest to O, the point Q farthest from O, and any other point R. (In Fig. A4.2 the line OR is tangent to k, but this need not be so; *any* point on k would do.) Thus the segment PQ is a diameter of k. We now look at the images P', Q', and R' of these points. We have $OP' = 1/OP$, $OQ' = 1/OQ$, and $OR' = 1/OR$. Therefore,

$$OP \cdot OP' = OQ \cdot OQ' = OR \cdot OR' = 1 \qquad (3)$$

or

$$\frac{OP}{OR} = \frac{OR'}{OP'}, \qquad \frac{OQ}{OR} = \frac{OR'}{OQ'} \qquad (4)$$

Also, the angle $\angle POR$ is common to the triangles $\triangle OPR$ and $\triangle OR'P'$. Thus these triangles are similar. Likewise, triangles $\triangle OQR$ and $\triangle OR'Q'$ are similar. Suppose now that R moves around the circle k in a clockwise direction. Then the angle $\angle PRQ$, subtending the diameter PQ, is a right angle. (Note that we have measured this angle in a clockwise sense, from P through R to Q.) We wish to show that the angle $\angle P'R'Q'$ is also a right angle, but measured in a counterclockwise sense.

To show this, we use a second theorem from elementary geometry: An external angle in a triangle equals the sum of the two internal angles not adjacent to it (Fig. A4.3). Thus $\angle OQ'R' = \angle OP'R' + \angle P'R'Q'$, or

$$\angle P'R'Q' = \angle OQ'R' - \angle OP'R' \qquad (5)$$

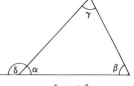

$\delta = \gamma + \beta$

Figure A 4.3

But since triangles $\triangle OPR$ and $\triangle OR'P'$ are similar, as are triangles $\triangle OQR$ and $\triangle OR'Q'$, we have $\angle OQ'R' = \angle ORQ$ and $\angle OP'R' = \angle ORP$. Thus Equation (5) becomes

$$\angle P'R'Q' = \angle ORQ - \angle ORP = \angle PRQ = 90°. \qquad (6)$$

Thus $\angle P'R'Q'$ is $90°$. Therefore as R moves around the circle k with diameter PQ, R' describes a circle k' with diameter $P'Q'$. This completes the proof.

241

Note that whereas the angle $\angle PRQ$ was measured clockwise, its "image angle" $\angle P'R'Q'$ is measured *counterclockwise;* this means that as the point R moves around k in a clockwise direction, its image point R' moves around k' in a *counterclockwise* direction. Inversion, like reflection, always reverses the sense of circular motion.

It is also interesting to note that although inversion carries the circle k to the circle k', the center of k is *not* carried to the center of k'; that is, although inversion preserves the property of being a circle, it does not preserve the property of being the center of the circle. This is because "being the center" is a property involving distance (the center is equidistant from all points of the circle), and as we have seen, inversion does not preserve distance.

The proofs given here are so-called "synthetic" proofs—they rely entirely on geometric constructions. One can prove the same properties using "analytic" proofs, i.e., the methods of analytic geometry. This has the advantage of efficiency—the same proof will establish at once Properties 1 and 2.[2]

A third property of inversion, regarding the preservation of angles, will be mentioned in connection with the stereographic projection.

5. Some Properties of the Stereographic Projection

We will prove three properties of the stereographic projection mentioned in the text. We assume that our sphere has a unit *diameter* (i.e., radius $1/2$; this will ensure that the image of the equator will be the unit circle in the plane). Let the sphere touch the plane of the map at the south pole S. Figure A5.1 shows a

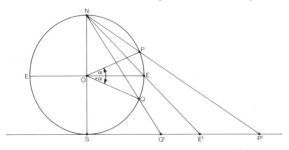

Figure A 5.1

[2] See *Elementary Mathematics from an Advanced Standpoint: Geometry* by Felix Klein, Dover Publications, New York, pp. 98–102.

cross section of the sphere; the segment *EE* represents the equator.

PROPERTY 1. *The stereographic projection maps two points having the same longitude but opposite latitudes onto two mutually inversive points on the map. In other words: Reflection in the equatorial plane of the sphere corresponds to inversion in the plane of the map.*

Proof: Let the two points be P and Q, with latitudes α and $-\alpha$, respectively. We now use a theorem from elementary geometry: An angle on the circumference of a circle equals one-half the central angle subtending the same arc (Fig. A5.2). Thus we have (Fig. A5.1):

$\angle ENP = \alpha/2$, since $\angle ENP$ and $\angle EOP$ subtend the same arc EP,
$\angle ENQ = -\alpha/2$, since $\angle ENQ$ and $\angle EOQ$ subtend the same arc EQ,
$\angle ENS = 45°$, since $\angle ENS$ and $\angle EOS$ subtend the same arc ES.

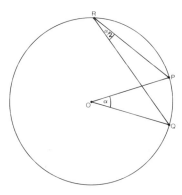

Figure A 5.2

Therefore $\angle PNS = 45° + \alpha/2$ and $\angle QNS = 45° - \alpha/2$. We now use the trigonometric function *tangent* (written "tan") to find the lengths of the segments SQ', SE', and SP'; in doing so we remember that $NS = 1$:

$$SQ' = \tan(45° - \alpha/2), \quad SE' = \tan 45° = 1, \quad SP' = \tan(45° + \alpha/2)$$

Using two well-known trigonometric identities (the so-called "sum and difference formulas"), the first and third of these equations can be written as

$$SQ' = \frac{1 - \tan \alpha/2}{1 + \tan \alpha/2}, \quad SP' = \frac{1 + \tan \alpha/2}{1 - \tan \alpha/2}$$

From this it follows that $SP' \cdot SQ' = 1$, showing that the image points P' and Q' are inverses of each other with respect to the equatorial circle on the map.

243

PROPERTY 2. *The stereographic projection maps every circle on the sphere onto a circle or a straight line on the map, and conversely.*

Proof: We will prove only the second part of this property, the one regarding straight lines. For the first part, the reader is referred elsewhere.[3]

Let the straight line in the plane of the map be *l*. We will project each point on *l* back onto the sphere by connecting it with a straight line to the north pole *N* (Fig. A5.3). The bundle

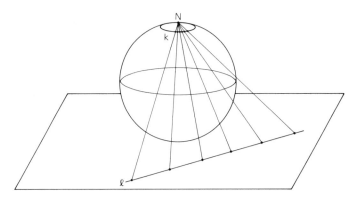

Figure A 5.3

of straight lines so created all lie in one plane, namely, the plane determined by *l* and *N*. This plane cuts the sphere in a circle *k* which passes through *N*. Thus every straight line in the plane of the map (including the lines emanating from *S*) is mapped onto a circle that passes through the north pole on the sphere.

To prove the converse of this statement, we use the symmetry property of our projection: If *P'* is the image of *P,* then *P* is the image of *P'*. Thus, as in inversion, the words "point" and "image point" can always be interchanged. It follows that every circle through the north pole on the sphere is carried to a straight line on the map.

PROPERTY 3. *The stereographic projection is conformal (angle-preserving).*

We will prove that if two curves in the plane of the map intersect at an angle α, their image curves on the sphere also intersect at the angle α; in other words, the projection is angle-preserving.

Proof: First we have to clarify what is meant by the angle of intersection of two curves. By this we mean the angle at which the *tangent lines* to the two curves meet at the point of intersection

---

[3] See *Elements of the Theory of Functions* by Konrad Knopp, translated by Frederick Bagemihl, Dover Publications, New York, 1952, pp. 35–39.

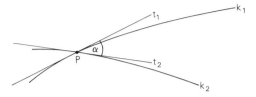

(Fig. A5.4). Thus in a small neighborhood of the point of intersection P, we may approximate the two curves k_1 and k_2 by their tangent lines t_1 and t_2.

We now project these tangent lines onto the sphere. Each will go over to a circle through the north pole; call these circles c_1 and c_2. Now c_1 and c_2 intersect at *two* points on the sphere—at N and at P', the image of P under the projection. Moreover, their angle of intersection is the same at N and P', as follows from the geometry of circles. But at N the tangent lines to c_1 and c_2 are horizontal and therefore parallel (in space) to the lines t_1 and t_2. Thus the angle of intersection of c_1 and c_2 at N, and therefore also at P', is equal to the angle of intersection of the original curves at P. This completes the proof.

Incidentally, the conformal property of the stereographic projection implies that inversion, too, is conformal. This follows from the fact that inversion in the plane corresponds to reflection in the equatorial plane of the sphere (Property 1). Since reflection, as well as the projection from the plane to the sphere, preserves angles, we see that the angle of intersection of two curves is preserved under inversion. (Note, however, that since a reflection always reverses the sense of rotation, only the magnitude of the angle is preserved; its sense is reversed.) Thus two perpendicular lines not passing through the center of inversion 0 go over to a pair of *orthogonal circles* (circles intersecting at right angles), both passing through 0. We have also seen how a hyperbola is carried over by inversion to the little eight-like figure centered at 0 (Fig. 12–5c). The two loops intersect at right angles as they should, since the two branches of the hyperbola "meet" at the point at infinity and are perpendicular there. Thus inversion, through its conformal property, actually enables us to see what goes on at infinity!

6. A Proof That There Exist Only Five Regular Solids

The proof is based on Euler's formula relating the number of faces F, the number of edges E, and the number of vertices V

of any simply connected polyhedron (a polyhedron having no holes):

$$V - E + F = 2 \qquad (1)$$

Suppose that our polyhedron has F identical faces, each being a regular polygon of n sides. (For example, for the cube we have $F = 6$, $n = 4$.) Then, counting the total number of edges, we have

$$nF = 2E \qquad (2)$$

since each edge belongs to two faces and is thus counted twice in the product nF. Suppose further that r edges meet at each vertex V. (For the cube, $V = 8$ and $r = 3$.) Then, counting again the number of edges, we have

$$rV = 2E \qquad (3)$$

since each edge connects two vertices. Substituting F and V from Equations (2) and (3) into Equation (1), we get

$$\frac{2E}{r} - E + \frac{2E}{n} = 2$$

or

$$\frac{1}{n} + \frac{1}{r} = \frac{1}{E} + \frac{1}{2} \qquad (4)$$

Now we know that $n \geq 3$ and $r \geq 3$, since a polygon must have at least three sides, and at least three edges must meet at each vertex of a polyhedron. But Equation (4) implies that n and r cannot *both* be greater than 3, for otherwise the left side of Equation (4) would be less than (or equal to) $1/2$, making E a negative number (or undefined if $n = r = 4$). Thus we only have to find the possible values of r when $n = 3$, and the possible values of n when $r = 3$:

For $n = 3$, Equation (4) becomes

$$\frac{1}{E} = \frac{1}{r} + \frac{1}{3} - \frac{1}{2} = \frac{1}{r} - \frac{1}{6}$$

Thus the possible values of r are 3, 4, or 5 ($r = 6$ would make E undefined; any value greater than 6 would make it negative). The corresponding values of E are 6, 12, or 30, giving, respectively, the tetrahedron, the octahedron, and the icosahedron.

Similarly, for $r = 3$ we get $n = 3$, 4, or 5. [Note that Equation (4) is symmetric in n and r, which means that these variables can be interchanged.] These values again correspond to $E = 6$,

12, or 30, but this time they give the tetrahedron, the cube, and the dodecahedron. (Note that these solids are the duals of the ones obtained for $n = 3$—a result of the symmetry just mentioned; see p. 105). These cases exhaust all possibilities. Thus, unlike the infinitely many regular polygons in the plane, there are just five regular solids in space.

7. The Concept of Group

A *group* is any collection of objects or "elements"—their precise nature is irrelevant—which fulfill the following four requirements:

1. Among the group elements an operation is defined so that when performing this operation on any two elements of the group, the result is always another element of the group. For want of a better name, it is customary to call this operation "multiplication" (although the group elements may not be numbers). Thus the result of "multiplying" the two elements a and b is their product, denoted ab. Since ab must always be an element of the group, we say that the group elements are *closed* under the multiplication.
2. Among the group elements there must be one, called the *identity* element and denoted by e, such that $ae = a$ for every a in the group; that is, the effect of multiplying any element by e is "to do nothing."
3. For every element a in the group there must be another element b, also in the group, such that $ab = e$; b is called the *inverse* of a.
4. The *associative rule* must hold for all elements of the group. That is, if a, b, and c are any elements of the group, we must always have $a(bc) = (ab)c$, so that the order of grouping the elements is immaterial.

Note that the *commutative* rule is not among the group requirements: it may well happen that $ab \neq ba$.

Following are several examples of groups:

1. The integers under addition. Here e is the number 0, and the inverse of an integer is its negative. (Indeed, $a + 0 = a$ and $a + (-a) = 0$ for any integer a.) Since the sum of two integers is always another integer, condition 1 (closure) is fulfilled. We also know, of course, that the addition of numbers is associative.
2. The rational numbers, with zero excluded, under multiplication. Here $e = 1$, and the inverse of a number a is its reciprocal $1/a$. Again closure and the associative rule are fulfilled, as we know

247

from the properties of rational numbers.

3. The set of numbers $\{1, -1, i, -i\}$, where $i^2 = -1$, under multiplication. This is an example of a *finite* group (the previous groups had infinitely many elements). Note that the subset $\{1, -1\}$ of this group is itself a group under the same operation, since all four requirements are fulfilled for this set. (This is not true for the set $\{i, -i\}$, since this set is not closed under multiplication; for instance, $i \cdot i = -1$; also it does not have an identity element.) Thus the set $\{1, -1\}$ is a *subgroup* of the larger group. The "multiplication table" for the entire group is shown below.

	1	*−1*	*i*	*−i*
1	*1*	*−1*	*i*	*−i*
−1	*−1*	*1*	*−i*	*i*
i	*i*	*−i*	*−1*	*1*
−i	*−i*	*i*	*1*	*−1*

4. An example of an *abstract* group—a group whose elements are not numbers—is the set of all vectors in the plane. A *vector* is a quantity having a magnitude and a direction; it is represented by an arrow or a directed line segment and is denoted by boldface letters, where the first letter is the initial point of the vector and the second letter is the end point. (Examples of vectors are translation, velocity, and force.) The group operation here is vector addition, which is done according to the familiar "triangle rule": to add the vectors **AB** and **CD** (Fig. A7.1), we first move **CD**

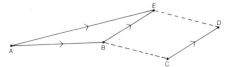

Figure A 7.1. *The addition of vectors.*

parallel to itself until its initial point C coincides with the end point B of **AB**, giving the vector **BE**. (Note that the vectors **CD** and **BE** are equal, since they have the same magnitude and direction.) Then the sum **AB** + **BE** is obtained by connecting the initial point A of **AB** with the end point E of **BE**, resulting in the vector **AE**. The *zero vector* is any vector of the form **AA**, i.e., a single point. The inverse of the vector **AB** is the vector **BA**—the two vectors have the same magnitude but opposite directions. (Indeed, **AB** + **BA** = **AA**, which is requirement no. 3.) Finally, we can show that the associative rule is always fulfilled: **AB** + (**BC** + **CD**) = (**AB** + **BC**) + **CD**. Indeed, the left-hand side of this equation, by the triangle rule, is **AB** + **BD**, which

in turn equals **AD,** while the right-hand side is equal to **AC** +
CD, or again **AD,** which proves the rule.

Since translations are vectors, we see that the set of all translations in the plane form a group. It is this group which we mentioned in Chapter 20 in connection with the symmetry elements
of a figure.

It can also be shown that the set of six symmetry elements of
an equilateral triangle (see p. 156) is a group. The group operation
here is the *combination* of two symmetry elements, i.e., their successive application. (In fact, this is an example of a non-commutative
group.)[4]

As a final note we mention two results about the combinations
of two reflections. If we combine two reflections in parallel mirrors,
the result is a *translation* across the mirrors through a distance
equal to twice the separation between them (Fig. A7.2). This is

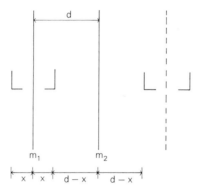

Figure A 7.2. *Two
reflections in parallel
mirrors result in a
translation across the
mirrors through twice the
separation of the mirrors.
(Note that* x + x + (d −
x) + (d − x) = 2d.)

the familiar sight at the barbershop, where you see the back of
your head in every *other* mirror. Similarly, the result of combining
two reflections in a pair of inclined mirrors (as in the kaleidoscope)
is a *rotation* through twice the angle between the mirrors, the
center of rotation being the point of intersection of the mirrors
(Fig. A7.3). We can see this when standing between two mirrors
inclined at 90° to each other: we see our reflected image in each
of the mirrors, but when looking towards their line of intersection,
we see ourselves as someone else would see us—our true image
turned through 180°. These two results have a bearing on the

---

[4] For details, see *Prelude to Mathematics* by W. W. Sawyer, Penguin Books,
Harmondsworth, 1966.

Figure A 7.3. *Two reflections in inclined mirrors result in a rotation through twice their angle of inclination. (Note that $\theta + \theta + (\alpha - \theta) + (\alpha - \theta) = 2\alpha$.)*

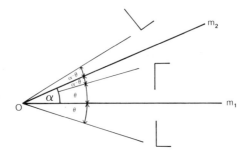

fact that the number of possible symmetry groups (along a line as well as in the plane) is finite.

8. Cantor's Proof That a Power Set Has More Elements than the Set Generating It

In the following discussion we will use the terminology of modern set theory: The number of elements of a set, finite or infinite, is its *cardinality* (Cantor had originally used the word *power*), while the set of all subsets of a given set is its *power set*.

Let the set be S and its power set S'. We wish to show that S' always has more elements than S; i.e., that S' has a greater cardinality than S.

For a finite set this is quite easy to show. If S has n elements, then S' will have 2^n elements, as can easily be shown by induction. Since $2^n > n$ for all $n = 0, 1, 2, \ldots$, we have proved our assertion.

For *infinite* sets, however, this proof is meaningless, since we have not assigned a meaning to the expression 2^n when n is infinite. Therefore Cantor had to base his proof entirely on the concept of *one-to-one correspondence;* he showed that such a correspondence *cannot* be established between S and S'. By the nature of things, his proof is somewhat abstract, since no use could be made of the formulas of ordinary (i.e., finite) algebra.

Assume that such a $1:1$ correspondence *does* exist. This means that for every element s in S there exists one, and only one, element s' in S', and vice versa; symbolically, $s \leftrightarrow s'$. Let us call the element s whose "image" is s' the "generator" of s'. (Of course, in view of the $1:1$ nature of the correspondence, the words "generator" and "image" can always be interchanged.) It should be pointed out that the actual rule which assigns this

250

correspondence can be quite arbitrary, as long as the $1:1$ requirement is met. To illustrate, suppose $S = \{a, b, c\}$. Then $S' = \{\{a\},$ $\{b\}, \{c\}, \{a, b\}, \{a, c\}, \{b, c\}, \{a, b, c\}, \{ \ \}\}$. (Note that we have included S itself, as well as the empty set $\{ \ \}$, in S'.) Now let our correspondence be as follows: $a \rightarrow \{a, b\}$, $b \rightarrow \{b, c\}$, $c \rightarrow \{a\}$. (We have used the symbol $\rightarrow$, rather than $\leftrightarrow$, since we actually know that S and S' cannot be put in a $1:1$ correspondence.)

Now the elements of S' can be classified into two classes: those elements (i.e., subsets of S) which contain their own generators, and those which don't. In our example above, each of the subsets $\{a, b\}$ and $\{b, c\}$ belong to the first class, since by our rule of correspondence the first of these sets has the generator a and the second has the generator b. However, the subset $\{a\}$ belongs to the second class, since its generator c is not contained in it.

We now form a new set T consisting of all those elements of S whose images belong to the second class in S'; that is, all the elements of S which are *not* contained in their own images. Since T consists of elements of S, it is itself a subset of S, and therefore is an element of S'. And since there exists a $1:1$ correspondence between S and S', T must have a generator in S. Call this generator s^*.

We now ask: Is s^* contained in T? If it is, this would contradict the definition of T as the set of all elements of S which are *not* contained in their own images. So s^* is not contained in T. But this again contradicts the definition of T, since T is the set of *all* elements of S which are not contained in their images. Thus s^* at the same time must and must not be contained in T, which obviously is impossible. Therefore, the assumption that a $1:1$ correspondence between S and S' can be established is false.

Thus we have shown that a set and its power set must have different cardinalities. To establish which one has the greater cardinality, we note that each element s of S can always be matched with one element of S' through the trivial correspondence $s \rightarrow \{s\}$. Thus every element in S has an image in S', but not every element in S' has a generator in S. Therefore S' has more elements than S: The cardinality of a power set is always greater than the cardinality of its "parent" set.

9. Some Recent Developments in Set Theory

Let us briefly survey the developments in set theory that have occurred since Cantor's pioneering work in the 1880s. For a fuller discussion the reader is referred to the extensive literature that exists on this subject.[5]

The first of these developments began with Cantor himself. In 1891 he thought of applying the ideas of set theory to the largest set we can think of—the *set of all sets*. Such an all-embracing concept would be a logical extension of his earlier work, but immediately he ran into a difficulty: What should be the cardinal number of this set? Obviously it must have the greatest possible cardinality of all sets. And yet, as Cantor had shown before, from any set one can create a new set with a larger cardinal number, namely, the power set of the original set. Thus the set of all sets *cannot* have the largest possible cardinal number. This turned out to be only the first of many paradoxes, or *antinomies,* involving the very definition of the set concept.

In 1902, the British mathematician and philosopher Bertrand Russell (1872–1970) came up with another paradox. Among the various sets we can think of, there are some with the peculiar property that they belong to themselves. For instance, let S be the set of *all objects describable with exactly eight English words.* Since S itself requires eight words for its description, it belongs to itself. Now consider the set R of all sets which do *not* belong to themselves. Does R belong to itself? If it does, this would contradict the definition of R. But if it does not, then it must belong to itself, because R was defined as the set of *all* sets which do not belong to themselves. So R must at the same time belong and not belong to itself, which is obviously impossible. In 1918 Russell gave a more popular version of this paradox, known as the Barber's Paradox. In a small village there was but one barber, and on his door was a sign which read: "I will shave anyone in this town who does not shave himself." In small print the sign added: "Offer is void for those who shave themselves." To his dismay, the barber one day discovered that he cannot fulfill his promise, for the following reason: Should he shave himself? If he does, he would violate his own promise to shave only those in town who don't shave themselves. But if he does not shave himself, then again he would violate his promise, since according to the promise he *must* shave himself!

Many more similar paradoxes have been discovered since. "I am a liar!", confesses a person in a moment of weakness. Is he

---

[5] See, for instance, the article "Non-Cantorian Set Theory" by Paul J. Cohen and Reuben Hersh, *Scientific American,* December 1967. An excellent survey of the development of mathematics in the nineteenth and twentieth centuries can be found in the book *Mathematics: The Loss of Certainty* by Morris Kline, Oxford University Press, New York, 1980.

telling the truth? If he does, then his own confession is a lie, which means that he is not a liar; but if he is not a liar, then his confession is true, so he *is* a liar! Or consider the pair of statements:

THE FOLLOWING SENTENCE IS FALSE
THE PRECEDING SENTENCE IS TRUE

Taken separately, each sentence is a valid statement; taken combined, they form a self-contradictory vicious circle, a kind of logical equivalent to Escher's impossible graphical loops.[6]

Paradoxes such as these led several mathematicians to reexamine the foundations of set theory. There was a growing feeling that this theory, successful as it was in describing infinite sets, cannot be based on Cantor's intuitive definition of a set as "any collection into a whole M of definite and separate objects m of our intuition or our thought." In particular, as Russell pointed out, this definition can bring us into difficulties when it is applied indiscriminately to such sets as the set of all sets. Russell insisted that in defining any particular set of objects, one should not use the very same set in the definition. In Russell's own words, "Whatever involves *all* of a collection must not be one of the collection." This would at once eliminate many of the paradoxes mentioned above.

But even this restriction was not enough to free set theory entirely from such logical loopholes. In an attempt to put the theory on a firm, rigorous basis, the German mathematician Ernst Zermelo (1871–1953) in 1908 formulated a system of nine axioms which define the concept of set and regulate its usage, in much the same way as the ten axioms of Euclid define the basic concepts of geometry and regulate their use. Specifically, the nine axioms define the basic relations between two sets (equality, union, and subset) and assure the existence of the empty ("null") set, of infinite sets, and of sets whose elements are themselves sets. Later, in 1922, Abraham A. Fränkel (1891–1965)[7] improved Zermelo's

---

[6] Yet another famous antinomy is the statement, "Every rule has an exception." If true, then this statement, too, must have an exception; that is, there must be some rules that do *not* have exceptions. But this makes the statement false! Thus the statement is self-contradictory, and seems to shake one of the most fundamental principles of logic, the so-called *principle of the excluded middle,* which says that every mathematical statement must be either true or false.

[7] Abraham (or in its Hebrew version, Avraham Halevi) Fränkel was my teacher at the Hebrew University of Jerusalem in 1956. His superb mathematical lectures, which he mingled with numerous anecdotes, made him a legendary figure among his many students and colleagues. I am glad to be able to pay him tribute in this book.

253

system of axioms, which has since been known as the Zermelo–Fränkel (ZF) system and has become the working basis for most of the research in set theory.

Among the nine ZF axioms there is one which, by its very nature, has assumed something of a special status and aroused a considerable amount of controversy. This is the celebrated *Axiom of Choice* (the eighth in the system), which says: *If S is a set (finite or infinite) of non-empty sets, and if no two sets in S have a common element, then it is possible to create a new set, the "choice set," consisting of precisely one element from each of the sets in S.*

For finite sets, this axiom is quite obvious: we merely have to select one element from each of the member sets of S, and this will immediately give us the choice set. But for *infinite* sets, a question arises: exactly how are we going to make our choice? This question may seem trivial, but remember, we now deal with an infinite number of sets, and experience has taught us that we cannot take things for granted when the infinite is involved. Indeed, we have already seen a statement whose truth seemed perfectly obvious, and yet whose subtle involvement with infinity has given mathematicians reason to question their own fundamental premises. This was Euclid's Fifth Postulate, the famous Parallel Postulate. The Axiom of Choice has assumed a similar status in set theory. Many mathematicians have felt that it cannot be taken for granted, that it is not as self-evident as, say, the first of the ZF axioms, which defines the equality of two sets ("two sets are equal if and only if they have the same elements").

The Axiom of Choice is important because the proofs of many theorems in set theory depend on its acceptance, just as many theorems in Euclidean geometry depend on accepting the Fifth Postulate (e.g., the theorem that the sum of the angles in a triangle is $180°$). It is for this reason that mathematicians were reluctant to dispense with this axiom. Rather than dispense with it, they first tried to prove it from the remaining ZF axioms, but these attempts were shown to be futile: first, the Austrian-American mathematician Kurt Gödel (1906–1978), in 1938, proved that the Axiom of Choice is consistent with the rest of the ZF axioms; that is, it cannot be disproved from them. Then in 1963, Paul Cohen (b. 1934) proved that not only cannot the Axiom of Choice be disproved from the other eight ZF axioms, it also cannot be *proved* from them. In other words, the axiom is independent of the rest of the axioms of set theory. Furthermore, Gödel and Cohen proved that the *Continuum Hypothesis* (Cantor's conjecture that there is no transfinite cardinal number between $\aleph_0$ and C; see p. 65) also can be neither proved nor disproved from the

254

ZF axioms (with or without the Axiom of Choice). Thus the famous problem with which Hilbert challenged the mathematical community in his address in 1900 was finally solved, but in a way which neither he nor his contemporaries could have ever anticipated.

Thus for the second time in its history, the concept of infinity has changed the course of mathematics. The realization that the Fifth Postulate is independent of the other Euclidean axioms meant that it could be replaced by an alternative, non-equivalent axiom. This led to the creation of non-Euclidean geometry, and in fact not just to one, but to *several* such geometries, depending on which alternative axiom is being chosen. Similarly, the discovery that the Axiom of Choice is logically independent of the other ZF axioms has given rise to several *non-Cantorian* set theories. The analogy between these two developments is even more striking because Gödel and Cohen used in their proofs "models" of sets in which the Axiom of Choice is replaced by various alternative anxioms, just as the nineteenth century mathematicians had used geometric models in which the Fifth Postulate was replaced by alternative axioms (we recall, for instance, the surface of the sphere, on which there are no parallel lines). These models really amount to new *interpretations* of known structures in which the new system of axioms holds (e.g., the surface of the Euclidean sphere is interpreted as a non-Euclidean plane).

By their very nature, the models used in non-Cantorian set theory are vastly more abstract than their non-Euclidean counterparts, and there is no easy way to "visualize" them. Nevertheless, their effect has been to demonstrate once again the *relative* nature of mathematics. Ever since the times of Thales and Pythagoras, mathematics has been hailed as the science of absolute and unfailing truth; its dictums were revered as the model of authority, its results trusted with absolute confidence. "In mathematics, an answer must be either true or false" is an age-old saying, and it reflects the high esteem which layman and professional alike have had for this discipline. The nineteenth century has put an end to this myth. As Gauss, Lobachevsky, and Bolyai have shown, there exist several different geometries, each of which is equally "true" from a logical standpoint. Which of these geometries we accept is a matter of choice, and depends solely on the premises (axioms) we agree upon. In our own century Gödel and Cohen showed that the same is true of set theory. But the implications go far beyond the realm of set theory. Since most mathematicians agree that set theory is the foundation upon which the entire structure of mathematics must be erected, the new discoveries amount to the realization that there is not just one, but several different mathematics, perhaps justifying the plural "s" with which the word has been used for centuries.[8]

Will these developments, so abstract in their nature, ever find any "practical" applications in the real world? No one can tell, of course; but if past events should serve as a guide, the answer may well turn out to be "yes." The history of mathematics is replete with discoveries which at first seemed to be totally abstract, but which later turned out to be of the utmost value to other sciences. We have seen how non-Euclidean geometry was at first accepted as a purely theoretical creation, but later found its way into the general theory of relativity. An even more dramatic example is that of group theory, a branch of algebra which only a century ago was regarded as the most abstract of all mathematical creations, and which today has become an indispensable tool in almost every branch of physics. These examples show that the course of mathematics, as of any science, is quite unpredictable, and one should never dismiss the possibility that some obscure branch of it may suddenly come to the forefront. As the physicist Niels Bohr (1885–1962) once said, "It is very hard to predict—especially the future!"

---

[8] It is interesting that just ten years before Gödel published his work on the Axiom of Choice, a discovery of similar consequences was made in physics. In 1927 Werner Heisenberg (1901–1976) enunciated his celebrated Uncertainty Principle, according to which one cannot determine simultaneously both the position and the momentum (i.e., the velocity) of a material particle, and this state of affairs is true regardless of the precision of the measuring device one uses. (It is instead a result of the wave–particle duality, the ability to describe a material object either as a particle or as a wave.) Heisenberg's principle laid to rest the classical notion that quantities such as position, velocity, and acceleration could, at least in principle, be determined with unlimited precision. This marked the end of determinism in the physical sciences, just as Gödel's theorem put an end to determinism in mathematics.

Bibliography

The following bibliography is necessarily incomplete. Especially in the field of cosmology there have appeared in recent years numerous books about the past and future of the universe, and it is next to impossible to list them all.

Part I

1. Petr Beckmann, *A History of PI,* The Golem Press, Colorado, 1977. A fascinating history not only of π but also of mathematics as part of our culture.
2. Bernard Bolzano, *Paradoxes of the Infinite,* Routledge & Kegan Paul, London, 1950 (originally published in 1851 and edited posthumously from the author's manuscript by Dr. Fr. Prihonsky). A short book of considerable historical significance, since it anticipated some of Cantor's ideas some 30 years before him.
3. Carl B. Boyer, *A History of Mathematics,* John Wiley, New York, 1968. A highly readable history of mathematics from the ancient Babylonians and Egyptians to the twentieth century.
4. Carl B. Boyer, *The History of Calculus and Its Conceptual Development,* Dover Publications, New York, 1959. This work includes an extensive discussion of the development of the ideas of continuity and limit.
5. Stephen I. Brown, *Some Prime Comparisons,* The National Council of Teachers of Mathematics, Reston, Virginia, 1978. Discusses the determination and distribution of primes. High school to college level.
6. Ronald Calinger, *Classics of Mathematics,* Moore Publishing Company, Oak Park, Illinois, 1982. Over a hundred short biographies of influential mathematicians, with excerpts from their writing.
7. Georg Cantor, *Contributions to the Founding of the Theory of Transfinite*

Numbers, Dover Publications, New York, 1955 (originally published in 1915), translated and provided with a historical introduction by Philip E.B. Jourdain. Here, in his own words, are Cantor's revolutionary ideas about the infinite. Except for the introduction, however, the treatment is highly technical.

8. Richard Courant and Herbert Robbins, *What Is Mathematics?,* Oxford University Press, London, 1969. An elementary introduction to higher mathematics. Discusses in detail topics from arithmetic, algebra, geometry, and calculus, many related to infinity.

9. Tobias Dantzig, *Number: The Language of Science,* The Free Press, New York, 1954. A cultural history of numbers and infinity.

10. Joseph Warren Dauben, *Georg Cantor: His Mathematics and Philosophy of the Infinite,* Harvard University Press, Cambridge, Massachusetts, 1979. The definitive biography of the creator of modern set theory.

11. Philip J. Davis, *The Lore of Large Numbers,* The Mathematical Association of America, Washington, D.C., 1961. The arithmetic of large numbers, discussed at the college level.

12. Howard Eves, *An Introduction to the History of Mathematics,* Holt, Rinehart and Winston, New York, 1976. A highly readable textbook in the history of mathematics, with hundreds of challenging exercises and their solutions.

13. Howard Eves and Carroll V. Newsom, *An Introduction to the Foundations and Fundamental Concepts of Mathematics,* Holt, Rinehart and Winston, New York, 1965. A college-level discussion of the foundations of mathematics, with chapters on Euclid's *Elements,* non-Euclidean geometry, sets, and logic.

14. George Gamow, *One, Two, Three . . . Infinity,* Viking, New York, 1948. A popular account of numbers and infinity.

15. Donald W. Hight, *A Concept of Limits,* Dover Publications, New York, 1977. A college-level discussion of limits.

16. L.B.J. Jolley, *Summation of Series,* Dover Publications, New York, 1961 (originally published in 1925). A collection of hundreds of finite and infinte series, classified according to their type. Highly enjoyable browsing for anyone fascinated by numbers.

17. E. Kamke, *Theory of Sets,* translated from the German by Frederick Bagemihl, Dover Publications, New York, 1950. A college-level discussion of modern set theory; pp. 1–51 discuss the arithmetic of infinity.

18. Morris Kline, *Mathematical Thought from Ancient to Modern Times,* Oxford University Press, New York, 1972. Chapter 20 of this exhaustive work is a history of infinite series.

19. Morris Kline, *Mathematics: The Loss of Certainty,* Oxford University Press, New York, 1980. A historical discussion of the impact that the discoveries of calculus, non-Euclidean geometry, and modern set theory had on the development of mathematics.

20. Edna E. Kramer, *The Nature and Growth of Modern Mathematics* (in two volumes), Fawcett Publications, Greenwich, Connecticut, 1974. A highly readable history of modern mathematics. No previous knowledge of mathematics is assumed.

258

21. J.R. Newman, *The World of Mathematics* (in four volumes), Simon and Schuster, New York, 1956. An anthology of writings in mathematics. Volume 3 contains essays by Bertrand Russell and Hans Hahn on infinity.

22. Rudy Rucker, *Infinity and the Mind: The Science and Philosophy of the Infinite,* Birkhäuser, Boston, 1982. A discussion of various mathematical and philosophical aspects of infinity, with particular emphasis on questions of logic.

23. Otto Toeplitz, *The Calculus: A Genetic Approach,* The University of Chicago Press, Chicago, 1963. This little book is a calculus textbook written from a historical viewpoint. The first chapter, "The Nature of the Infinite Process," is particularly revealing.

24. Leo Zippin, *Uses of Infinity,* The Mathematical Association of America, Washington, D.C., 1962. A college-level discussion of numbers and infinity.

Part II

1. *The Thirteen Books of Euclid's Elements* (in three volumes), translated from the text of Heiberg with introduction and commentary by Sir Thomas Heath, Dover Publications, New York, 1956. The classic and definitive account of Greek geometry. Next to the Bible, the *Elements* is believed to be the most widely translated book of all time.

2. Theodore Andrea Cook, *The Curves of Life,* Dover Publications, New York, 1979 (originally published in 1914). An exhaustive account of the logarithmic spiral in art and nature, with hundreds of photographs and illustrations.

3. H.S.M. Coxeter, *Introduction to Geometry,* John Wiley, New York, 1969. A college-level treatise of geometry, written from the standpoint of groups of transformations and symmetry.

4. Howard Eves, *A Survey of Geometry,* Allyn and Bacon, Boston, 1972. A college-level textbook, with an extensive discussion of Euclidean, non-Euclidean, and projective geometry, transformations, and more.

5. Matila Ghyka, *The Geometry of Art and Life,* Dover Publications, New York, 1977. An elementary discussion of various topics in geometry and their influence on the fine arts. Among the topics are proportions in space and time, the golden section, and the regular partitions of the plane and space. The text is accompanied by many illustrations and photographs.

6. Patrick Highes and George Brecht, *Vicious Circles and Infinity—A Panoply of Paradoxes,* Doubleday, New York, 1975 (originally published in 1939). An amusing collection of paradoxes, some relating to infinity, but none discussed in depth.

7. David Hilbert and S. Cohn-Vossen, *Geometry and the Imagination,* translated from the German by P. Nemenyi, Chelsea Publishing Company, New York, 1952 (originally published in 1932). An elementary—though far from simple—account of geometry, with extensive discussions of plane and space tessellations, projective and non-Euclidean geometry, topology, and more.

8. H.E. Huntley, *The Divine Proportion: A Study in Mathematical Beauty,* Dover Publications, New York, 1970. Various mathematical topics (the golden section, the logarithmic spiral, Pascal's triangle, and Fibonacci numbers) and their aesthetic meaning. Many illustrations.

9. Konrad Knopp, *Elements of the Theory of Functions,* translated from the German by Frederick Bagemihl, Dover Publications, New York, 1952. A classic in mathematics, discussing topics from complex numbers, circular inversion, and the theory of functions of complex variables. Chapter 3 is a discussion of the stereographic projection.

10. Benoit B. Mandelbrot, *The Fractal Geometry of Nature,* W.H. Freeman, San Francisco, 1982. A richly illustrated treatise of the new geometry of fractals (fractional dimensions), with particular emphasis on some of the so-called "pathological curves."

11. Albert E. Meder, *Topics from Inversive Geometry,* Houghton Mifflin, Boston, 1967. A beginning college-level discussion of circular inversion, using elementary analytic geometry.

12. Bruno Munari, *Discovery of the Circle,* George Wittenborn, New York, 1970. Really an art book, with hundreds of illustrations and photographs showing the circle in its various manifestations in art and geometry. There is also a companion book, *Discovery of the Square.*

13. Phares G. O'Daffer and Stanley R. Clemens, *Geometry: An Investigative Approach,* Addison-Wesley, Menlo Park, California, 1977. An elementary textbook with much material about plane and space tessellations, transformations, and symmetry.

14. Dan Pedoe, *Geometry and the Liberal Arts,* St. Martin's Press, New York, 1976. Discusses the influence of various topics in geometry on the visual arts, among them the golden section, perspective, and proportion.

15. Garth E. Runion, *The Golden Section and Related Curiosa,* Scott, Foresman and Company, Glenview, Illinois, 1972. A mathematical discussion of the golden section, the logarithmic spiral, and the Fibonacci numbers. High school to college level.

16. Robert C. Yates, *Curves and Their Properties,* The National Council of Teachers of Mathematics, Reston, Virginia, 1952. An encyclopedia of common and not-so-common plane curves, with their graphs and properties. Many of the curves are related to infinity.

Part III

1. F.H. Bool, J.R. Kist, J.L. Locher, and F. Wierda, *M.C. Escher: His Life and Complete Graphic Work,* Harry N. Abrams, New York, 1981. An extensive biography of Escher's life and work, with the full text of his book on tessellations. Includes excellent reproductions of all of Escher's prints.

2. Edwards B. Edwards, *Pattern and Design with Dynamic Symmetry,* Dover Publications, New York, 1967 (originally published in 1932). This little book contains numerous designs based on the logarithmic spiral, with a discussion of dynamic symmetry and its application in art.

3. Bruno Ernst, *The Magic Mirror of M.C. Escher,* Random House, New

York, 1976. This book, written by a mathematician who was a close friend of Escher, is an excellent introduction to the mathematical aspects of Escher's work.

4. E.H. Gombrich, *The Sense of Order: A Study in the Psychology of the Decorative Arts,* Cornell University Press, Ithaca, New York, 1979. This thorough work covers practically every aspect of the decorative arts, with hundreds of illustrations and photographs.

5. J.L. Locher, *The World of M.C. Escher,* Harry N. Abrams, New York, 1971. Contains most of Escher's graphic work, some in color, as well as several articles on his work, including his own essay, "Approaches to Infinity."

6. Marjorie Hope Nicolson, *Mountain Gloom and Mountain Glory: The Development of the Aesthetics of the Infinite,* W.W. Norton, New York, 1959. Discusses how the great discoveries of the seventeenth century in physics and astronomy changed man's perception of nature, space, and time. Includes many poems and literary comments on infinity.

7. Robert Sietsema, *Designs of the Ancient World,* Hart Publishing Company, New York, 1978. A collection of graphic designs and patterns from various ancient civilizations.

8. Peter S. Stevens, *Handbook of Regular Pattern: An Introduction to Symmetry in Two Dimensions,* The MIT Press, Cambridge, Massachusetts, 1981. A lavishly illustrated encyclopedia of regular patterns, classified according to the symmetry groups to which they belong.

9. Hermann Weyl, *Symmetry,* Princeton University Press, Princeton, New Jersey, 1952. The classic treatise on the mathematical and aesthetic aspects of symmetry, with a semi-technical discussion of symmetry groups. Richly illustrated.

Part IV

1. Emile Borel, *Space & Time,* Dover Publications, New York, 1960. A popular, though somewhat outdated, exposition of space, time, and the special and general theories of relativity. Particularly revealing is the discussion of the relations between mathematical and physical space.

2. Paul Davies, *The Edge of Infinity,* Simon and Schuster, New York, 1981. A popular account of recent developments in cosmology— black holes, singularities, and the Big Bang.

3. Sir Arthur Eddington, *The Expanding Universe,* The University of Michigan Press, Ann Arbor, Michigan, 1958 (originally published in 1932). This little work, written by one of the great astronomers of the early twentieth century, is a non-technical account of the recession of the galaxies. Of course, much of the discussion is long outdated, but it still provides charming reading.

4. Timothy Ferris, *To the Red Limit: The Search for the Edge of the Universe,* Bantam Books, New York, 1979. A popular account of black holes, quasars, and the expanding universe, with a fascinating portrait of the scientists behind these discoveries.

5. Timothy Ferris, *Galaxies,* Sierra Club, New York, 1982. A spectacular

photographic album of the universe, concentrating on galaxies and clusters of galaxies.

6. Alexander Koyŕe, *From the Closed World to the Infinite Universe,* Johns Hopkins University Press, Baltimore, 1974. A historical discussion of the emergence of the concept of an infinite universe after Copernicus.

7. Philip Morrison, Phylis Morrison and the Office of Charles and Ray Eames, *Powers of Ten,* Scientific American Library, San Francisco, 1982. A photographic tour of the universe, arranged according to the order of magnitude of its objects from the subnuclear particles to the entire universe.

8. Milton K. Munitz, *Theories of the Universe from Babylonian Myth to Modern Science,* The Free Press, Glencoe, Illinois, 1957. An anthology of writings about the structure of the universe from ancient to modern times.

9. Dorothea Waley Singer, *Giordano Bruno: His Life and Thought,* with an annotated translation of his work, "On the Infinite Universe and Worlds," Henry Schuman, New York, 1950. A biography of the man who sacrificed his life for believing in an infinite universe.

10. Steven Weinberg, *The First Three Minutes: A Modern View of the Origin of the Universe,* Basic Books, New York, 1976. A popular and highly readable account of the events immediately following the Big Bang. There is a fascinating description of the discovery of the three-degree microwave radiation that is the remnant from the Big Bang.

11. Louise B. Young, *The Mystery of Matter,* Oxford University Press, New York, 1965. An anthology of writings about the structure of matter, the atomistic theory, and the question of whether matter is infinitely divisible.

Index